7th International Conference on Applied Physics and Mathematics (ICAPM 2017)

Journal of Physics: Conference Series Volume 814

Tokyo, Japan
20 – 22 January 2017

ISBN: 978-1-5108-3859-8
ISSN: 1742-6588

Printed from e-media with permission by:

Curran Associates, Inc.
57 Morehouse Lane
Red Hook, NY 12571

Some format issues inherent in the e-media version may also appear in this print version.

Printed by Curran Associates, Inc. (2017)

Institute of Physics
Dirac House, Temple Back
Bristol BS1 6BE UK

Phone: 44 1 17 929 7481
Fax: 44 1 17 920 0979

techtracking@iop.org

Additional copies of this publication are available from:

Curran Associates, Inc.
57 Morehouse Lane
Red Hook, NY 12571 USA
Phone: 845-758-0400
Fax: 845-758-2633
Email: curran@proceedings.com
Web: www.proceedings.com

Table of contents

Volume 814

7th International Conference on Applied Physics and Mathematics (ICAPM 2017)
20–22 January 2017, Tokyo, Japan

Accepted papers received: 16 February 2017
Published online: 31 March 2017

Preface

011001
OPEN ACCESS
7th International Conference on Applied Physics and Mathematics (ICAPM 2017)

011002
OPEN ACCESS
Conference Committee

011003
OPEN ACCESS
Peer review statement

Papers

Mathematical Modeling and Analysis

012001
OPEN ACCESS
Evaluation of lead-time procurement in Pupuk Kaltim Inc. using pure-birth analysis

Novaldy Pratama Putra and Atina Ahdika.....1

012002
OPEN ACCESS
Comparison of INAR(1)-Poisson model and Markov prediction model in forecasting the number of DHF patients in west java Indonesia

Atina Ahdika and Novyan Lusiyana.....8

012003
OPEN ACCESS
Numerical simulations on the flow fields of dynamic axial compression columns in chromatography processes

Ru Chien Liang, Cheng Che Liu, Ming Tsai Liang and Jiann Lin Chen.....15

012004
OPEN ACCESS
Numerical study of MPS method with large eddy simulation for fluid solid coupling problem

Chao YANG, Huaixin ZHANG and Huilan YAO.....22

012005
OPEN ACCESS
First order transition in two dimensional coulomb glass

Preeti Bhandari and Vikas Malik.....27

012006
OPEN ACCESS
A model for inter-laminar shear stress in laminated composites

W Zaki, V Nguyen and R Umer.....33

012007
OPEN ACCESS
Comparison between a typical and a simplified model for blast load-induced structural response

A Abd-Elhamed and S Mahmoud.....37

Computer and Mathematical Engineering

012008
OPEN ACCESS
The research of 4th grade mathematical curriculum electronic picture book construction and development in integrating indigenous culture

Yen Ting Chen and Juei Hsin Wang.....45

012009
OPEN ACCESS
Almost Boolean fuzzy rings

B A Alaba and Y G Wondifraw.....52

012010
OPEN ACCESS
Efficient large scale commute time embedding

H I Hahn.....58

012011
OPEN ACCESS
Comparison of combustion products by the iteration method and application of gasoline and biogas

Paramust Juntarakod.....63

012012
OPEN ACCESS
Integrated optimisation technique based on computer-aided capacity and safety evaluation for managing downstream lane-drop merging area of signalised junctions

CHAI Chen and WONG Yiik Diew.....72

Applied Mathematics

012013
OPEN ACCESS
Some results on circulant and skew circulant type matrices with k-Fibonacci sequences

Ying Zhang.....79

012014
OPEN ACCESS
A new spectral method using legendre wavelets for shallow water model in limited-area

Fukang Yin, Junqiang Song, Jianping Wu and Xiaoqun Cao.....88

012015
OPEN ACCESS
Finite time quenching for a nonlinear diffusion equation with singular boundary flux

Ying Yang.....96

012016
OPEN ACCESS
Rational Chebyshev pseudospectral method for long-short wave equations

Zeting Liu and Shujuan Lv.....102

012017
OPEN ACCESS
An accelerated randomized extended Kaczmarz algorithm

Xu Xiang, Xu Liu, Wentang Tan and Xiang Dai.....108

012018
OPEN ACCESS
Economic growth and concentrations

Shu-hen Chiang.....116

012019
OPEN ACCESS
Approximate solutions for the nonlinear duffing oscillator with an external force

Jutamas Kirinin, Kanokwan Mahiththummath, Narisara Kongsawasdi and Jaipong Kasemsuwan.....122

Physical Electronics and Applications

012020
OPEN ACCESS
Application of exponential homotopy algorithm in the inverter harmonic elimination

Li Zhang.....129

012021
OPEN ACCESS
Non-orthogonal frequency shift keying for optical wireless link based on visible light emitting diode

Yong-Yuk Won, Sang Min Yoon and Dongsun Seo.....135

012022
OPEN ACCESS
Improved synchronization criteria for time-delayed chaotic Lur'e systems using sampled-data control

Wenyong Duan, Yan Li, Xiaorong Fu and Baozhu Du.....139

012023
OPEN ACCESS
Efficient parallel analysis of transient plane wave scattering from nonplanar doubly periodic structures

Nan-Wei Chen.....144

PREFACE

It is our great pleasure to welcome you to 2017 7th International Conference on Applied Physics and Mathematics (ICAPM 2017) which will be held in Tokyo, Japan during Jan. 20-22, 2017. ICAPM2017 is dedicated to issues related to Applied Physics and Mathematics.

The major goal and feature of the conference is to bring academic scientists, engineers, industry researchers together to exchange and share their experiences and research results, and discuss the practical challenges encountered and the solutions adopted. Professors from Japan and Taiwan are invited to deliver keynote speeches regarding latest information in their respective expertise areas. It will be a golden opportunity for the students, researchers and engineers to interact with the experts and specialists to get their advice or consultation on technical matters, sales and marketing strategies.

This proceedings present a selection from papers submitted to the conference from universities, research institutes and industries. All of the papers were subjected to peer-review by conference committee members and international reviewers. The papers selected depended on their quality and their relevancy to the conference. The volume tends to present to the readers the recent advances in the field of Applied Physics and Mathematics and various related areas, such as Applied Mathematics and Engineering: Modeling, Analysis and Computation, Mathematical methods in optics and electromagnetics, Nonlinear problems in mechanics, Applications of Microscopy in the Physical Sciences, Nanoscience and Nanotechnology, Applied Non-linear Physics, etc..

We would like to thank all the authors who have contributed to this volume and also to the organizing committee, reviewers, speakers, chairpersons, sponsors and all the conference participants for their support to ICAPM 2017.

Prof. Naoyuki Ishimura
Chuo University, Japan
Jan. 12, 2017

IOP Conf. Series: Journal of Physics: Conf. Series **814** (2017) 011001 doi:10.1088/1742-6596/814/1/011001

Statement of Peer Review

All papers published in this volume of Journal of Physics: Conference Series have been peer reviewed through processes administered by the proceedings Editors. Reviews were conducted by expert referees to the professional and scientific standards expected of a proceedings journal published by IOP Publishing.

IOP Conf. Series: Journal of Physics: Conf. Series **814** (2017) 011002 doi:10.1088/1742-6596/814/1/011002

Conference Committee

Conference Chair

Prof. Naoyuki Ishimura, Chuo University, Japan

Prof. Po-Liang Liu, National Chung-Hsing University, Taiwan

Program Chair

Prof. Vadim R. Gasiyarov, South Ural State University, Russia

Prof. Yoshihiko Uematsu,Gifu University, Japan

Prof. David Greenhalgh, University of Strathclyde, UK

Technical Committee

Assoc. Prof. Wael Zaki, Khalifa University, UAE

Prof. Dongsun Seo, Myongji University, South Korea

Asst. Prof. Yong-Yuk Won, Dept. of Electronic Engineering, Myongji University, South Korea

Assoc. Prof. Jiann Lin Chen, I-Shou University, Taiwan

Asst. Prof. Dr. Vikas Malik, Jaypee Institute of Information Technology, India

Prof. Hee-Il Hahn, Hankuk University of Foreign Studies, Korea

Asst. Prof. Dr. Yan-Gu Lin, National Synchrotron Radiation Research Center, Taiwan

Prof. Nikolay Sokolov, Russian Academy of Sciences, Moscow, Russia

Prof. Alexander Balinsky, Cardiff University, UK

Prof. Dr. Abdur Rashid, Gomal University, Pakistan

Prof. Nan-Wei Chen, Yuan Ze University, Taiwan

Prof. Roberto Montemanni, University of Applied Sciences of Southern Switzerland, Switzerland

Assoc. Prof. Kuo-Chih Chuang, Zhejiang University, China

Assoc. Prof. Yen Ting Chen, National Taichung University of Education, Taiwan

Prof. Juei Hsin Wang, National Taichung University of Education, Taiwan

Prof. Dr. Shurong Zhou, School of Life Sciences, Fudan University, China

Dr. Ayman Abd-Elhamed, Faculty of Engineering at Mataria, Helwan University, Egypt

Dr. Khushboo Verma, UNIVERSITY OF LUCKNOW, India

Dr. Wooi-Nee Tan, Multimedia University, Malaysia

Assoc. Prof. Praveen Agarwal, Anand Intl. College of Engg, India

Assoc. Prof. Dr. Feng Xiao, Tokyo Institute of Technology, Yokohama, Japan

Prof. Marat Akhmet, Middle East Technical University, Turkey

Assoc. Prof. Dr. Dinshaw Balsara, University of Notre Dame, Notre Dame, Indiana, USA

Prof. Haydar Akca, College of Arts and Sciences, Abu Dhabi University, UAE

Asst. Prof. Thamir Abdul Ameer Hassan, Al-Karkh University of Science Iraq, Iraq

Prof. Xiaodi Li, School of Mathematical Sciences, Shandong Normal University, China

Prof. Abdelaziz Soufyane, ALHOSN University, Abu Dhabi, UAE

Assoc. Prof. S. Joseph Antony, University of Leeds, UK

Prof. Steven B Damelin, the University of Michigan, Ann Arbor, USA

Prof. Rosa Pardo, Complutense University of Madrid, Spain

Peer review statement

All papers published in this volume of *Journal of Physics: Conference Series* have been peer reviewed through processes administered by the proceedings Editors. Reviews were conducted by expert referees to the professional and scientific standards expected of a proceedings journal published by IOP Publishing.

Evaluation of lead-time procurement in Pupuk Kaltim Inc. using pure-birth analysis

Novaldy Pratama Putra[1] and Atina Ahdika[2]

[1]Student at Department of Statistics, Universitas Islam Indonesia, Jalan Kaliurang Km 14.5 Sleman, Yogyakarta, 55584, Indonesia
[2]Lecturer at Department of Statistics, Universitas Islam Indonesia, Jalan Kaliurang Km 14.5 Sleman, Yogyakarta, 55584, Indonesia

atina.a@uii.ac.id

Abstract. Industry has an important role in supporting national economy by its contributions to the employment absorption and creating added value of commodities produced. One of the parameters that influence the success of an industry is how the industry evaluates its performance, in particular, the performance of service procurement department. A good industry is an industry which can provide excellent service and well-timed. In the case study, we use pure-birth analysis to evaluate the lead-time procurement in Pupuk Kaltim Inc by collecting the service duration time in each process to be followed in the procurement department and modeling its probability to finish each work at any time. The result shows that the greatest probability of the division to finish their work is around 80 days. Based on the result, we conclude that they have to improve their service performance.

1. Introduction

Economic development is a multi-dimensional process. This process consists not only by transformation of economic structure but also by economic composition [1] Industry is one of the strategic economic activities to increase public revenues and the economic development is indicated by the increase of employment, technology transfer and the increase of foreign exchange.
In practice, many industrial systems consist of multiple units. If one unit fails then others unit will be fail too. It means that failures has statistically correlated and depend on the previous condition. There are three types failure interaction ([2]; [3]; [4]):

1. Type I failure interaction: When a unit fails, other units will immediately fail at a certain probability.
2. Type II failure interaction: The failure of a unit is regarded as an internal shock that increases failure rates of other units.
3. Type III failure interaction: A failure of one unit causes a random amount of damage to the other unit.

Pupuk Kaltim Inc is a State-Owned Enterprises in Indonesia, which was established on December 7th 1977 with the main objectives are to implement and support the policy of the Government in terms of industrial development and national economy, particularly the industrial sector of the fertilizer and chemical industries. The main reason we decide Pupuk Kaltim Inc as our research object is because they are the biggest fertilizer producer to support agriculture sector in Indonesia. As we know,

agriculture is one of main occupations of Indonesian people. Department that most responsible for meeting the needs of the service in any part of Pupuk Kaltim Inc is the service procurement department. One of the most important parameters which can decide the success of the department tasks is duration service procurement itself. In fact, more than 50% procurement in 2015 is out of the targeted time.
The procurement process consists of four stages, that are the determination method and list proposed partner (DUR), clarification of the negotiation, the report of winner tender (LUP), and the last is the manufacture of the letter of agreement (SP). All these stages must always be sequentially, it means that the stage-two could not be done before the stage-one is completed, and so on. One of the most appropriate methods to analyze a case like this is pure birth analysis in Continuous Time Markov Chain (CMTC).

2. Stochastic process

Markov Chain analysis is one of the statistical methods which can be used to estimate the future in probabilistic point of view. The analysis works by using information from the past to obtain the information of the future. Markov Chain is a special form of probabilistic model known as stochastic process [5]. There are two parameters involved in Markov Chain analysis; state and time parameter. Based on the type of the data, the state and time parameters can be discrete or continuous type. Therefore, there are four types of Markov Chain model. In this paper, we built a Markov Chain model with discrete state parameters and continuous time parameters. The model is known as Continuous Time Markov Chain (CTMC). CTMC is a stochastic process having Markov property; conditional distribution of $X(t+s)$ in the future given by $X(s)$ in the present and $X(u)$ from the past, $0 \le u < s$, is only determined by the present and independent from the past and can be notated as [6]

$$P\left(X\left(t+s\right)=j \mid X\left(s\right)=i\right)$$

In other words, CTMC is a stochastic process moving from a state to another that is discrete, however the amount of time spent in each state before move to another state is exponentially distributed. CTMC can be described as a birth and death process, where the process which have states $\{0,1,2,...\}$ can move only from state i to state $i+1$ or to state $i-1$. The relationship among the states is shown by figure 1.

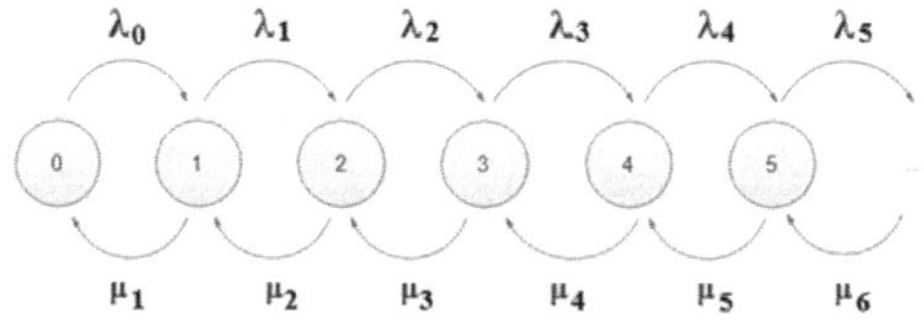

Figure 1. Diagram of Birth Death Process

Where the transition rates are

$$q_{i,j} = \begin{cases} \lambda_i & if \quad j=i+1 \\ \mu_i & if \quad j=i-1 \\ 0 & others \end{cases}$$

The transition rate $q_{i,j}$ can be written in matrix generator Q form as follow

$$Q = \begin{pmatrix} q_{0,0} & q_{0,1} & q_{0,2} & \cdots & q_{0,n} \\ q_{1,0} & q_{1,1} & q_{1,2} & \cdots & q_{1,n} \\ q_{2,0} & q_{2.1} & q_{2,2} & \cdots & q_{2,n} \\ \vdots & & & \ddots & \vdots \\ q_{n,0} & q_{n,1} & q_{n,2} & \cdots & q_{n,n} \end{pmatrix} = \begin{pmatrix} -\lambda_0 & \lambda_0 & 0 & \cdots & 0 \\ \mu_1 & -(\lambda_1+\mu_1) & \lambda_1 & \cdots & 0 \\ 0 & \mu_2 & -(\lambda_2+\mu_2) & \cdots & 0 \\ \vdots & & & \ddots & \vdots \\ 0 & 0 & 0 & & \end{pmatrix} \quad (1)$$

From matrix Q, we can obtain time-dependent solution of probability in state n by solving the global balance condition

$$\frac{d}{dt}\pi(t) = \pi(t)\cdot Q \tag{2}$$

Where $\pi(t) = [\pi_0(t) \;\; \pi_1(t) \;\; \pi_2(t) \;\; \cdots]$

3. Analysis and discussion

There are four main processes in the procurement process, namely the determination method and list proposed partner (DUR), clarification of the negotiation, reports of winner tender (LUP), and the last is the manufacture of the letter of agreement (SP). In the DUR process contains a list of vendors who have capability on tender required by the user. In this process involves two to three persons namely Manager Procurement and General Manager, and if the value of services up to more than 50 million, it needs the approval of the Director.

The second process is a negotiation process, this process is a process that most fluctuated than the other processes because there are various problems encountered. This process is carried out by the department of procurement and vendor in order to get the deal price.

The third process is the process LUP or Proposed Statement winner. After getting the price of the winning bidder, the next process is the winner Proposed Statement. This process was signed by the three person on this is the team leader, GM unit, and directors.

The process of the fourth or the last process is the SP or the Letter of Agreement. In this process involves six parties, namely the procurement manger, General Manager of Engineering, Team Lead, User Manager, User Manager and Directors General. Of the four main processes are then distributed according to the number of people who sign it so that the letter obtained 12 processes. All these processes will be built into models of stochastic processes. Construction of this stochastic process models through the following steps, namely formulate postulates that satisfies Poisson and Markov processes, model postulates to the form of differential equations (PD) and the use of partial differential equations theorem for generating functions of transition opportunities, as well as the form and function of a joint moment.

To get a differential equation it is necessary to calculate the average of time duration of the respective data, where the states of the system and the person who is sign the contract are shown in Table 1:

Table 1. List of states and the person in charge

State	Process	Person in Charge
0	DUR (List of Business Partner)	Manager of Service Procurement
1		General Manager
2		Technology and Development Director
3	Negotiation	-
4	LUP (Report of the Tender's Winner)	Team Leader
5		Project General Manager
6		Director
7	SP (Memorandum of Understanding)	Manager of Service Procurement
8		General Manager of Technology
9		Team Leader
10		User Manager
11		User General Manager
12		Director

Table 2. Duration time of the service procurement based on the states in the system

Tender	State (day)											
	0-1	1-2	2-3	3-4	4-5	5-6	6-7	7-8	8-9	9-10	10-11	11-12

1	4	7	4	101	0	0	14	0	0	0	0	14
2	0	3	1	15	0	2	3	0	2	0	2	1
3	0	3	1	58	0	5	7	0	12	0	6	1
4	0	3	0	28	1	2	27	1	2	1	2	27
⋮	⋮	⋮	⋮	⋮	⋮	⋮	⋮	⋮	⋮	⋮	⋮	⋮
47	0	0	1	11	0	1	15	0	1	0	1	15
# days/tender	0.12	2.04	1.14	20.00	3.45	2.78	9.09	1.23	0.12	2.38	4.35	3.70
# tenders/day	8.11	0.49	0.88	0.05	0.29	0.36	0.11	0.81	8.11	0.42	0.23	0.27

From the process shown in Table 2, it was found that the type of Markov chain is Pure-Birth Process. This is because all the processes occurred from t to $t + 1$ and there is no process that shows the reverse to the previous process. For more details, the series of processes can be described in figure 2

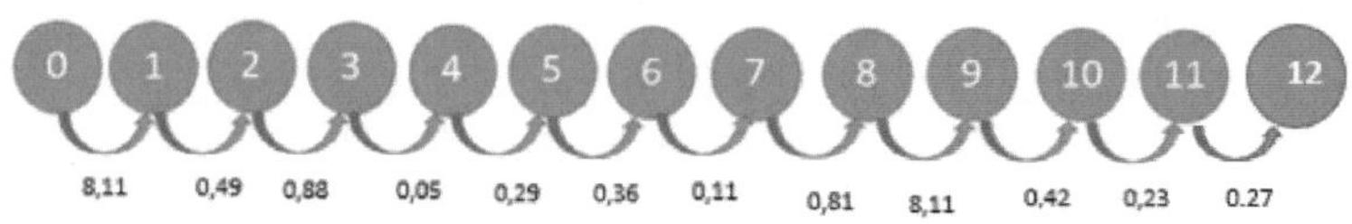

Figure 2. Diagram of pure-birth Process in service procurement system

The probability of the process will be in state i can be obtained by solving the differential equation using global balance condition (Eq.(2)). The differential equation for each process is shown in Table 3

Table 3. Differential equation for each process

State	Process	Differential Equation
0		$\pi'_0(t) = -8.11\pi_0(t)$
1	DUR (List of Business Partner)	$\pi'_1(t) = 8.11\pi_0(t) - 0.49\pi_1(t)$
2		$\pi'_2(t) = 0.49\pi_1(t) - 0.88\pi_2(t)$
3	Negotiation	$\pi'_3(t) = 0.88\pi_2(t) - 0.05\pi_3(t)$
4		$\pi'_4(t) = 0.05\pi_3(t) - 0.29\pi_4(t)$
5	LUP (Report of the Tender's Winner)	$\pi'_5(t) = 0.29\pi_4(t) - 0.36\pi_5(t)$
6		$\pi'_6(t) = 0.36\pi_5(t) - 0.11\pi_6(t)$
7		$\pi'_7(t) = 0.11\pi_6(t) - 0.81\pi_7(t)$
8		$\pi'_8(t) = 0.81\pi_7(t) - 8.11\pi_8(t)$
9	SP (Memorandum of Understanding)	$\pi'_9(t) = 8.11\pi_8(t) - 0.42\pi_9(t)$
10		$\pi'_{10}(t) = 0.42\pi_9(t) - 0.23\pi_{10}(t)$
11		$\pi'_{11}(t) = 0.23\pi_{10}(t) - 0.27\pi_{11}(t)$
12		$\pi'_{12}(t) = 0.27\pi_{11}(t)$

After getting the differential equations, then the next step is calculating the solution. Below are the probability distributions of the tender's completion in each process.

3.1. DUR process

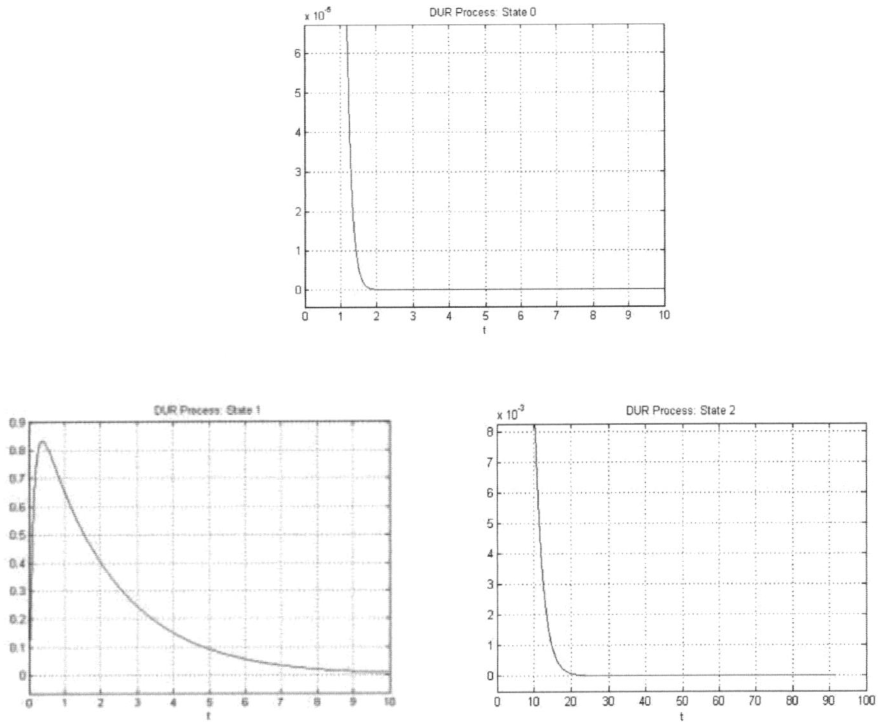

Figure 3. DUR Process in: State 0, State 1, and State 2

From figure 3, we can see that the List Proposed Partner (DUR) is conducted by the procurement manager (state 0), the team leader (state 1), and directors (state 2). In state 1, the greatest probability to complete the work is less than 1 day. Meanwhile, after two days the chances are zero. This occurs because the incoming mail right in the procurement department, after going through the secretary of the letter straight to the procurement manager. In state 1, the biggest probability that the work will be completed is also less than one day. While in state 2 the directors take up to 2 days to complete their works. Based on these results the greatest probability of completion of the work on DUR process is 4 days. This includes ideal because the total days to complete the work in DUR process is less than 10 days, suitable with the company's target.

3.2. Negotiation process

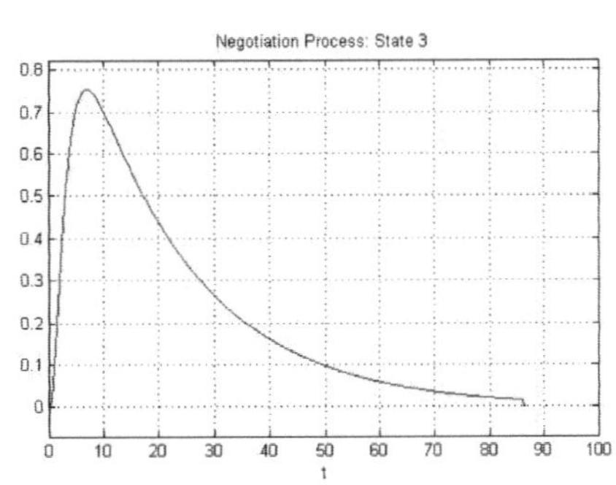

Figure 4. Negotiation Process

IOP Conf. Series: Journal of Physics: Conf. Series **814** (2017) 012001 doi:10.1088/1742-6596/814/1/012001

In the negotiation process, based on the result in figure 4, the biggest probability of the completion of the work is about seven days. When passing the seventh day, the probability of the work to be completed is decrease. This does not mean that the work will not be completed when passing the seventh day, but the process of negotiations on the method of appointment is mostly to be completed on the seventh day.

3.3. LUP process

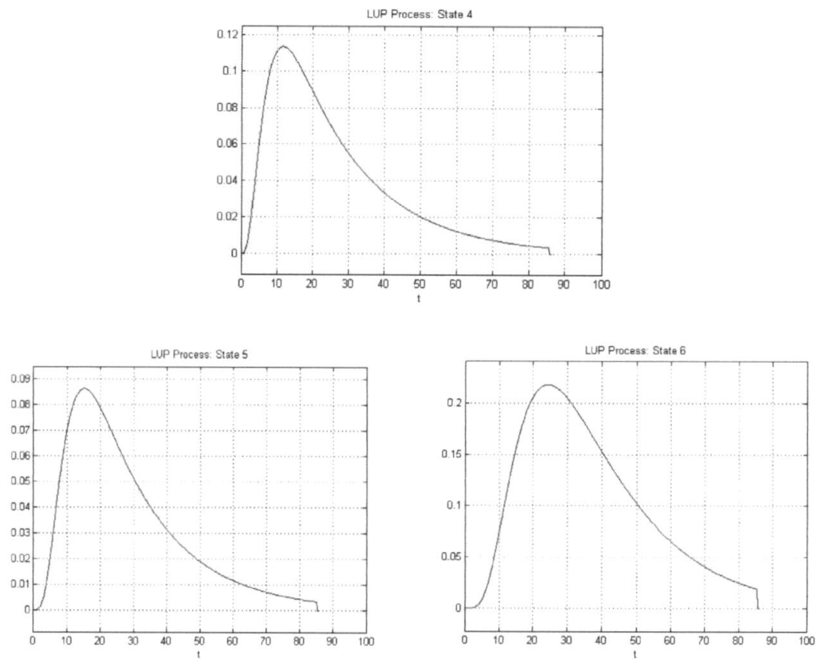

Figure 5. LUP Process in: State 4, State 5, and State 6

The process of LUP (Report Proposed winner) is conducted by team leader (state 4), project general manager (state 5), and director (state 6). From figure 5, we can see that in the LUP process the probability to complete the work is increase until reach about 15 to 25 days. State 4 is carried out by the team leader where greatest probability to finish the work is in 11th day. Then the next process is carried out by project general manager where the greatest probability to finish the work is 0.8 which is on the 15th day. While in the state 6 the work done by the director where the biggest probability to finish is 0.2 which is on the 20th day. So the total of days required is until about 20th day.

3.4. SP process

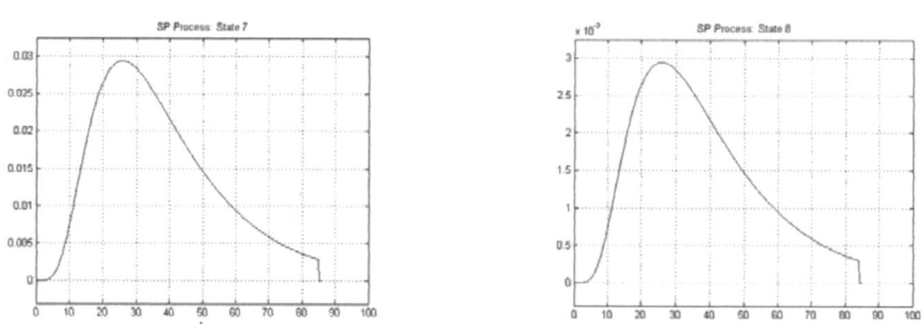

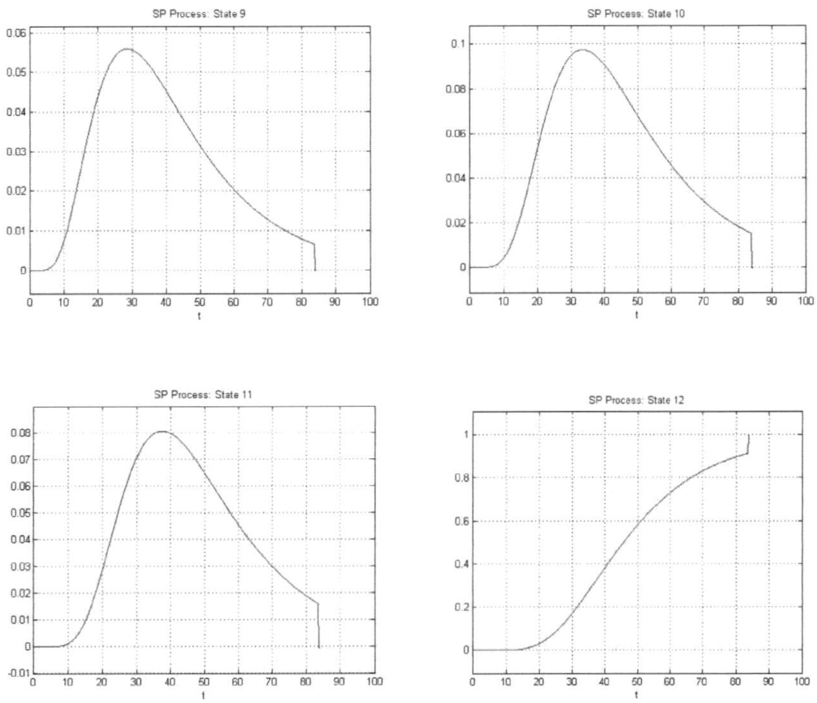

Figure 6. SP process in: state 7, state 8, state 9, state 10, state 11, and state 12

SP process is carried out by six parts: manager of service procurement, general manager of technology, team leader, user manager, user general manager, and directors. On state 7 and 8, the process is undertaken by the manager of service procurement and general manager of technology. From the result shown in figure 6, the biggest probability of the work's completion of this processes are at 25th day. In state 9, 10, and 11, the greatest probability to finish each work are on 30th, 35th, and 37th day. While in state 12 the largest probability to complete the work is on day 80 that is equal to 1.

4. Conclusion

We have constructed a stochastic model to evaluate the lead time procurement in Pupuk Kaltim Inc. using continuous time Markov chain especially pure birth process. From the analysis, we can conclude that the highest probability to finish the work in DUR process can be reached for about 1 day, in negotiation process is until about seventh day, in LUP process is until about 25th day, and in SP process is until about 80th day. So the total day required by each tender to be finished is about 80 days. Based on government's policy, it needs 60 days for a tender to be finished in a company.

5. References

[1] Todaro, Michael P. 2003. Economic Development in The Third World. Translated by: Aminuddin and Drs.Mursid. Jakarta: Ghalia Indonesia.
[2] Murthy DNP, and Wilson RJ. 1994. Parameter Estimation in Multi-Component with Failure Interaction. Stochastic Models and Data Analysis 1994; 10:47-60.
[3] Murthy DNP, and Nguyen DG. 1985. Study of Two-Component System with Failure Interaction. Naval Research Logistics 1985; 10: 239-247.
[4] Lai MT, and Chen YC. 2008. *Optimal Replacement Period of a Two-Unit System with Failure Rate Interaction and External Shocks.* International Journal of Systems Science 2008; 39: 71-79.
[5] Norris, James R. 1998. Markov Chains. Cambridge University Press. Retrieved 2016-03-04.
[6] Ross, SM. 2014. Introduction to Probability Models, 11th Edition. UK: Academic Press.

IOP Conf. Series: Journal of Physics: Conf. Series **814** (2017) 012002 doi:10.1088/1742-6596/814/1/012002

Comparison of INAR(1)-Poisson model and Markov prediction model in forecasting the number of DHF patients in west java Indonesia

Atina Ahdika[1] and Novyan Lusiyana[2]

[1]Department of Statistics, Faculty of Mathematics and Natural Sciences, Universitas Islam Indonesia, Jalan Kaliurang Km 14.5 Sleman, Yogyakarta, Indonesia 55584
[2]Department of Parasitology, Faculty of Medicine, Universitas Islam Indonesia, Jalan Kaliurang Km 14.5 Sleman, Yogyakarta, Indonesia 55584

atina.a@uii.ac.id

Abstract. World Health Organization (WHO) noted Indonesia as the country with the highest dengue (DHF) cases in Southeast Asia. There are no vaccine and specific treatment for DHF. One of the efforts which can be done by both government and resident is doing a prevention action. In statistics, there are some methods to predict the number of DHF cases to be used as the reference to prevent the DHF cases. In this paper, a discrete time series model, INAR(1)-Poisson model in specific, and Markov prediction model are used to predict the number of DHF patients in West Java Indonesia. The result shows that MPM is the best model since it has the smallest value of MAE (mean absolute error) and MAPE (mean absolute percentage error).

1. Introduction

Dengue hemorrhagic fever (DHF) is one of the major health problems in Indonesia. DHF is caused by dengue virus, a positive-strand RNA virus of flaviviridae family with four distinct serotypes (DEN1-4) that transmitted by *Aedes aegypti* (*Ae. aegypti*) and *Aedes albopictus* (*Ae. albopictus*) [1]. DHF were also endemic, occurred throughout the year and can be accompanied by extraordinary events [2].

World Health Organization (WHO) states that the prevention and reduction of dengue virus transmission was strongly influenced by control vector and control human contact with mosquitoes, because until now there were not found a vaccine and specific treatment for DHF [3-4]. Although the treatment for DHF is not found yet, the government should know the estimation of the next cases which possibly happened to do a prevention action. The number of reported cases at time t are usually influenced by the number of cases at time $t-1$. This kind of data can be predicted using time series model. Many researchers have estimated the number of DHF incidence using time series analysis. Promprou *et al.* (2006) used autoregressive integrated moving average (ARIMA) to model and forecast the monthly number of DHF cases in southern Thailand [5]. Choudhury *et al.* (2008) used seasonal autoregressive moving average (SARIMA) $(1,0,0)(1,1,1)_{12}$ to model the monthly number of DF cases in Dhaka, Bangladesh [6]. Martinez *et al.* (2011) used SARIMA $(2,1,2)(1,1,1)_{12}$ model to predict dengue incidence in Campinas, Southeast Brazil [7]. Sriwattanapongse and Khanabsakdi (2011) used additive and multiplicative regression models to describe the patterns of hospital-diagnosed Malaria and DHF incidences by using the previous monthly or quarterly periods of incidences occurring in the upper Northern region of Thailand [8].

Based on the previous researches, the purpose of this paper is to predict the number of DHF cases. The data we used are the number of DHF patients in a hospital in West Java Indonesia. Most of previous researches used continuous time series model to predict the number of DHF cases because their data range are quite wide. While the data used in this paper have smaller range, then the suitable time series model is discrete time series, specifically integer-valued autoregressive (INAR) order 1 model with Poisson marginal distribution. As a comparison, we also model the number of DHF patients using Markov prediction model (MPM). The MPM model is the probabilistic model which use the concept of the probability of an event occurred based on the previous event. The two models are compared using three kinds of measures; MSE, MAE, and MAPE.

The paper is organized as follows. In Section 2, we provide the theoretical basis of INAR(1)-Poisson. MPM model will be provided in Section 3. In Section 4, the measures of comparison for each model will be given. Then the data are modelled and analysed in Section 5 which divided into the modelling and comparing steps. In Section 6, we will draw a conclusion based on the analysis result.

2. INAR(1)-Poisson model

INAR(1) model is defined as follows [9]:

$$X_t = \alpha \circ X_{t-1} + \varepsilon_t, \quad t = 0, \pm 1, \pm 2, \ldots \tag{1}$$

where $\alpha \in [0,1]$ and ε_t is a sequence of uncorrelated non-negative integer-valued random variables having mean μ and finite variance σ^2 (follows any discrete distribution).

The INAR(1) model defined in Eq.(1) states that the total individuals at time t, X_t, are the summation of the survivors of the individuals at time $t-1$, X_{t-1}, each with probability of survival α and the individuals entered the system in interval $(t-1, t]$, ε_t.

Operator $'\circ'$ in Eq.(1) is defined as [10]:

$$\alpha \circ X_{t-1} = \sum_{i=1}^{X_{t-1}} W_i \tag{2}$$

where $W_i \sim Bernoulli(\alpha)$.

In this paper, the marginal distribution of ε_t is Poisson distribution, and the model so called INAR(1)-Poisson. Al-Osh and Alzaid (1987) state that X_t has a Poisson distribution if and only if ε_t has a Poisson distribution. If $\varepsilon_t \sim Poi(\lambda(1-\alpha))$, using probability generating function technique of Eqn.(1) and discrete self-decomposable properties [10], then $X_t \sim Poi(\lambda)$.

There are some techniques to estimates the parameters of INAR(1)-Poisson model, α and λ. In this paper, we estimate the parameters using Yule-Walker and conditional least square (CLS) estimation.

2.1 *Yule-Walker estimators*

The Yule-Walker estimators can be obtained by replacing the covariance at lag k, $\gamma(k)$, with the sample autocovariance function defined by [9]

$$\gamma(k) = \alpha^k \, var(X_{t-k}) + \sum_{j=0}^{k-1} \alpha^j \, cov(X_{t-k}, \varepsilon_{t-j}) = \alpha^k \, \gamma(0) \tag{3}$$

and solve for α to obtain

$$\hat{\alpha} = \frac{\sum_{t=0}^{n-1}(x_t - \bar{x})(x_{t+1} - \bar{x})}{\sum_{t=0}^{n}(x_t - \bar{x})^2} \tag{4}$$

where $\bar{x}$ is the sample mean. To estimate for λ, calculate $\hat{\varepsilon}_t = x_t - \hat{\alpha}\, x_{t-1}$ for $t = 1,2,\ldots,n$ where $X_t \sim Poi(\lambda)$ and obtained

$$\hat{\lambda} = \sum_{t=1}^{n} \hat{\varepsilon}_t / n \quad (5)$$

2.2 CLS estimators

The conditional mean of X_t given X_{t-1} is given by [9]

$$E(X_t|X_{t-1}) = \alpha X_{t-1} + \lambda = g(\theta, X_{t-1}) \quad (6)$$

where $\theta = (\alpha, \lambda)$ is the set of parameters to be estimated.
The CLS estimation is derived from the minimization of the sum of squared deviations of the conditional expectation in Eq.(6), that is find the value of α and λ which minimize

$$Q_n(\theta) = \sum_{t=1}^{n} [(X_t - g(\theta, X_{t-1})]^2 \quad (7)$$

with respect to θ. From the derivation, the estimators of α and λ are obtained

$$\hat{\alpha} = \frac{\sum x_t x_{t-1} - \frac{\sum x_t \sum x_{t-1}}{n}}{\sum x_{t-1}^2 - \frac{(\sum x_{t-1})^2}{n}} \quad (8)$$

and

$$\hat{\lambda} = \frac{1}{n}(\sum x_t - \hat{\alpha} \sum x_{t-1}) \quad (9)$$

3. Markov prediction model

Analogue with INAR(1)-Poisson model, Markov prediction model (MPM) can also be used to model the data at time $\boldsymbol{t}$ which depend on the data at time $\boldsymbol{t-1}$. Suppose $\{\boldsymbol{X(t)}, \boldsymbol{t} = \boldsymbol{0}, \boldsymbol{1}, \boldsymbol{2}, \dots\}$ is a stochastic process that takes on a finite number of possible value. If $\boldsymbol{X(t)} = \boldsymbol{i}$ means that the process is in state $\boldsymbol{i}$ at time $\boldsymbol{t}$, then we can calculate the probability of $\boldsymbol{X(t+1)} = \boldsymbol{j}|\boldsymbol{X(t)} = \boldsymbol{i}$, that is the probability that the process will be in state $\boldsymbol{j}$ at time $\boldsymbol{t+1}$ given the previous process is in state $\boldsymbol{i}$ at time $\boldsymbol{t}$. Suppose that

$$P(X(t+1) = j|X(t) = i, X(t-1) = i_{t-1}, \dots, X_1 = i_1, X_0 = i_0) = P_{ij} \quad (10)$$

for all state $i_0, i_1, \dots, i_{t-1}, i, j$ and all $n \geq 0$. These stochastic process is known as a Markov chain [11].

P_{ij} in Eq.(10) has the properties as follows

$$P_{ij} \geq 0, \quad i,j \geq 0; \quad \sum_{j=0}^{\infty} P_{ij} = 1, \quad i = 0,1,\dots$$

Let P denote the matrix of one-step transition probabilities P_{ij}, so that

$$P = \begin{pmatrix} P_{00} & P_{01} & P_{02} & \cdots \\ P_{10} & P_{11} & P_{12} & \cdots \\ \vdots & \vdots & \vdots & \end{pmatrix} \quad (11)$$

Then define the n −step transition probabilities P_{ij}^n to be the probability that the process in state i will be in state j after n additional transition. To compute the n −step transition probability, we can use Chapman-Kolmogorov equations,

$$P_{ij}^{n+m} = \sum_{k=0}^{\infty} P_{ik}^{n} P_{kj}^{m}, \quad for\ all\ n, m \geq 0, all\ i, j \tag{12}$$

Suppose $X(t) = [0 \ 1 \ 2 \ 3 \ ... \ c]$ is the set of the number of DHF patients, where c is the maximum number of DHF patients, analogues with the Markov prediction model which was developed by Ren *et al.* (2015) [12], the MPM for this case is as follows

$$\begin{aligned} X(1) &= X(0) \cdot P \\ X(2) &= X(1) \cdot P \\ &\vdots \\ X(t+1) &= X(t) \cdot P \end{aligned} \tag{13}$$

where P is the transition probability matrix defined by Eq.(11). Using Eq.(13), we can predict the number of DHF patients at time $t + 1$ given by the number of DHF patients at time t.

4. MSE, MAE, and MAPE

There are many kinds of measure which can be used to determine the best model of a certain data. In this paper, we will use mean squared error (MSE), mean absolute error (MAE), and mean absolute percentage error (MAPE) to compare and choose the best model. The formula of three measures are given by

$$MSE = \frac{1}{n}\sum_{t=1}^{n}(\hat{X}_t - X_t)^2 \tag{14}$$

$$MAE = \frac{1}{n}\sum_{t=1}^{n}|\hat{X}_t - X_t| \tag{15}$$

$$MAPE = \frac{1}{n}\sum_{t=1}^{n}\left|\frac{X_t - \hat{X}_t}{X_t}\right| \cdot 100 \tag{16}$$

5. Result and analysis

Data used in this paper is the number of DHF patients in a certain hospital in West Java Indonesia daily from December 27th 2014 to August 24th 2015. Suppose that $X(t)$ defines the number of DHF patients at time t. The description of the data is shown in figure 1

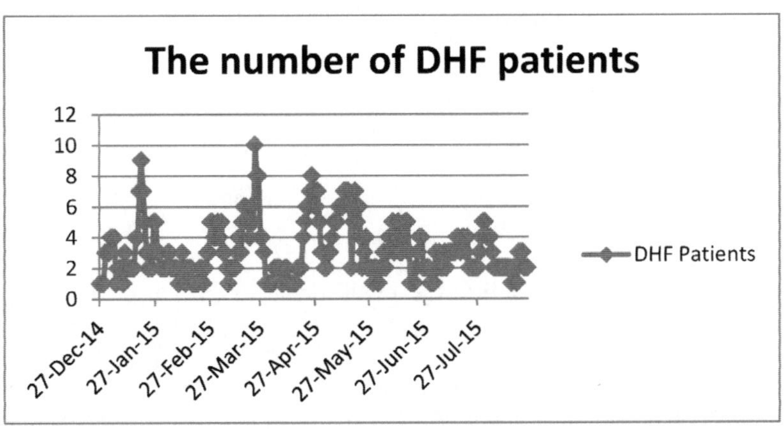

Figure 1. The number of DHF patients

From figure 1, we can see that the number of DHF patients in those hospital are in range $[1,10]$. Then we can define that $c = 10$. The maximum number of DHF patients is reached in March 24th 2015. First step to modeling a data using INAR(1)-Poisson is checking the distribution and PACF of the data. In testing the hypothesis of the data distribution, we use level of significance $\alpha = 5\%$. H_0 state that the data follows Poisson distribution and H_1 state that data did not follow Poisson distribution. H_0 will be rejected if the value of $\text{Sig.} < \alpha$. Based on KS test, we obtain the value of $\text{Sig.} = 0.076 > \alpha$, it means that H_0 did not rejected so that the data follows Poisson distribution. Figure 2 shows the partial autocorrelation function (PACF) to determine the order of the INAR model.

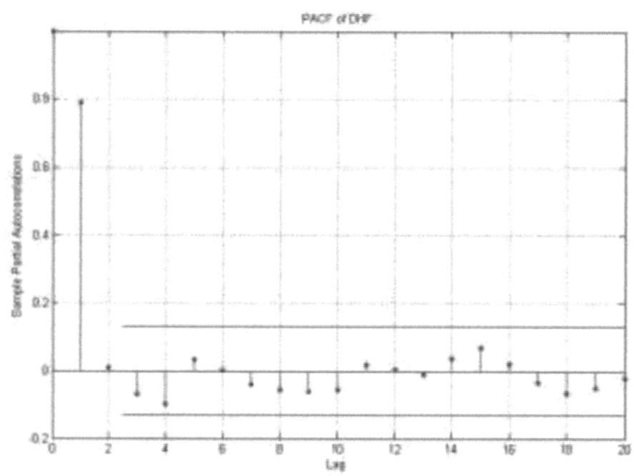

Figure 2. PACF of DHF patients

Figure 2 shows that the lag of the data of DHF patients is in lag 1, so the INAR model for this data is INAR order 1. Table 1 shows the estimation parameter of INAR(1)-Poisson model using CLS and Yule-Walker estimation.

Table 1. Parameter estimation of INAR(1) model

Parameters	CLS estimation	Yule-Walker estimation
$\hat{\alpha}$	0.7904	0.7892
$\hat{\lambda}$	0.6530	0.6542

Based on the estimation results, we have two INAR(1)-Poisson models that are

$$\text{Model 1: } X_t = 0.7904 \circ X_{t-1} + \varepsilon_t, \quad \varepsilon_t \sim Poi(0.6530) \tag{17}$$

$$\text{Model 2: } X_t = 0.7892 \circ X_{t-1} + \varepsilon_t, \quad \varepsilon_t \sim Poi(0.6542) \tag{18}$$

In predicting the number of DHF patients at time t given the number of patients at time $t-1$, we have to build the transition probability matrix P of the data. Suppose that $X(t)$ define the number of DHF patients at time t. Based on figure 1, it is known that the number of DHF patients is in range $[1,10]$ so we have 10 states. $P(X(t) = j|X(t-1) = i)$ is the probability that the number of DHF patients is j at time t given by the number of patients is i at time $t-1$. The transition probability matrix P obtained from the calculation is given by

$P =$

0.5143	0.3714	0.1143	0	0	0	0	0	0	0
0.1605	0.6173	0.1728	0.0247	0.0247	0	0	0	0	0

0.0444	0.3111	0.3333	0.2667	0.0444	0	0	0	0	0
0.0323	0.1290	0.1613	0.3548	0.2581	0.0323	0.0323	0	0	0
0	0	0.3043	0.1739	0.2609	0.1739	0.0435	0	0	0.0435
0	0.0909	0	0.0909	0.1818	0.3636	0.2727	0	0	0
0	0	0	0	0.3333	0.2222	0.2222	0.1111	0.1111	0
0	0	0	0.3333	0	0	0.3333	0.3333	0	0
0	0	0	0	0	0	1	0	0	0
0	0	0	0	0	0	0	1	0	0

The prediction of the number of DHF patients then being predicted using Eq.(13) where $X(t) = [1\ \ 2\ \ 3\ \ \ldots\ \ 10]$. Figure 3 shows the prediction of the number of DHF patients using INAR(1)-Poisson model with CLS and Yule-Walker estimation and using MPM.

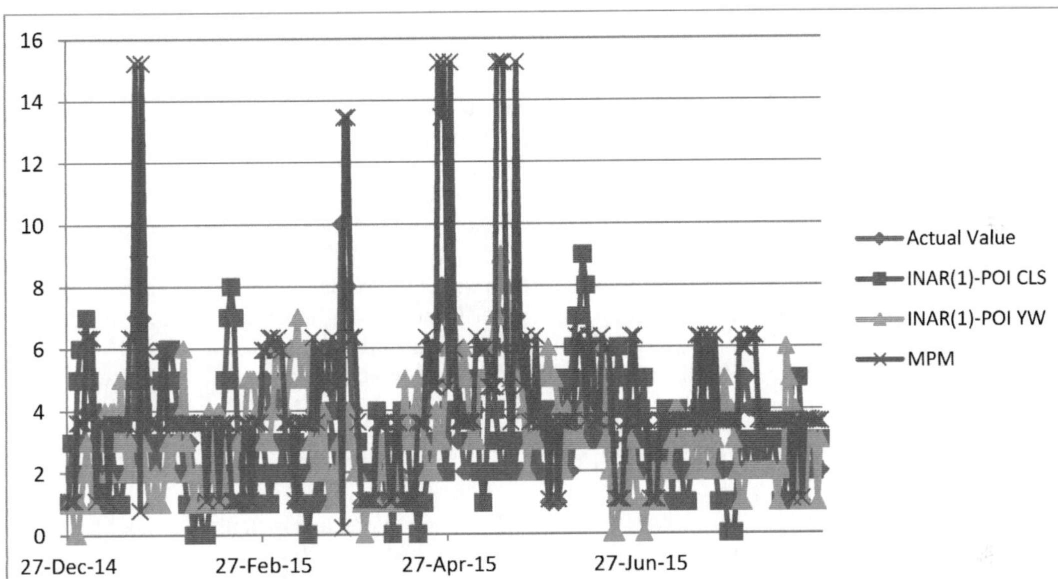

Figure 3. Prediction data using INAR(1)-Poisson with CLS and Yule-Walker estimation and using MPM compared to actual value

Figure 3 shows the comparison between the prediction data using INAR(1)-Poisson model and MPM and the actual value. From figure 3 we can see that most of the patterns of the prediction data are quite near with the actual value. In some points, the prediction data from the MPM have further distance compared to the other models. To choose the best model, Table 2 provides the value of the MSE, MAE, and MAPE of each model.

Table 2. MSE, MAE, and MAPE of each model

Model	MSE	MAE	MAPE
INAR(1)-Poisson CLS	4.784232365	1.705394191	70.03029704 %
INAR(1)-Poisson Yule-Walker	4.460580913	1.589211618	66.00342488 %
MPM	5.3760102	1.57491237	48.8855449 %

Based on the value of MSE, the best model to predict the number of DHF patients is INAR(1)-Poisson with Yule-Walker estimation. While using MAE and MAPE, the best model is MPM. In general, we can conclude that MPM has better result in predicting the number of DHF patients. The result is rather different if we compare it with the graphic in figure 3. It may happen because the prediction data of MPM which have further distance to the actual value may only a few points. While the other point are closer to the actual value. Otherwise, the INAR(1)-Poisson model may have more points which did not close with the actual data although the distance is not so far away.

6. Conclusion

We have compared the prediction model of the number of DHF patients in a hospital in West Java Indonesia using three models; INAR(1)-Poisson with CLS estimation, INAR(1)-Poisson with Yule-Walker estimation, and Markov prediction model (MPM). Based on the result, MPM has the smallest value of MAE and MAPE compared to the other models although it has the biggest value of MSE. We conclude that, in general, MPM is the best model to predict the number of DHF patients.

7. References

[1] Kumar VS, Mani P, Banstin TMM, Ravikumar G. 2012. Mosquito Larvicidal, Oviposition deterrent and Repellent properties of *Acalypha indica* L extracts against *Aedes aegypti*, *Anopheles stephensi*, and *Culex quinquefasciatus*. Int J Med Biosci 1(3):33–41.

[2] Lidia K, Setianingrum ELS. 2008. Deteksi dini resistensi nyamuk *Aedes albopictus* terhadap insektisida organofosfat di daerah endemis demam berdarah dengue di Palu (Sulawesi tengah). *MKM*. 3:2.

[3] Basker P., Kannan P., Porkaipandian R.T., Saravanan S. Sridharan S., and Khadiresan M. 2013. Study on entomological surveilance and its significance during a dengue outbreak in the district of Tirunelvi in Tamil Nandu, India. Osong Public Health Res Perspect. 2013.4(3):152-158.

[4] WHO. 1999. Demam berdarah dengue: diagnosis, pengobatan, pencegahan dan pengendalian, edisi 2. WHO.Genewa.

[5] Promprou, S., Jaroensutasinee, M., and Jaroensutasinee, K. 2006. Forecasting dengue haemorrhagic fever cases in southern Thailand using ARIMA models. Dengue Bulletin, Vol. 30, 2006.

[6] Choudhury, M.A.H.Z., Banu, S., and Islam, M.A. 2008. Forecasting dengue incidence in Dhaka, Bangladesh: A time series analysis. Dengue Bulletin, Vol. 32, 2008.

[7] Martinez, E.Z., da Silva, E.A.S., and Fabbro, A.L.D. 2011. A SARIMA forecasting model to predict the number of cases of dengue in Campinas, State of Sao Paulo, Brazil. Revista da Sociedade Brasileira de Medicina Tropical 44(4); 436-440, Jul-Ago, 2011.

[8] Sriwattanapongse, W. and Khanabsakdi, S. 2011. Modeling and forecasting malaria and dengue hemorrhagic fever incidence and prevalence in Northern Thailand. Journal of Mathematics and System Science I (2011) 52-59.

[9] Al-Osh, M.A., and Alzaid, A.A. 1987. First order integer valued autoregressive (INAR(1)) process. Journal of Time Series Analysis 8: 261-275.

[10] Steutel, F.W., and Harn, K.V. 1979. Discrete analogues of self-decomposability and stability. The Annals of Probability, Vol. 7, No. 5, 893-899.

[11] Ross, S.M. 2007. *Introduction to Probability Models, 9th Edition*. USA: Academic Press.

[12] Ren, X., Liu, Q., and Zhang, Y. 2015. The proportion of energy consumption structure prediction based on markov chain. Journal of Chemical and Pharmaceutical Research, 2015, 7(3): 378-382.

Acknowledgments

A sincere gratitude goes to the Academic Development Board of Universitas Islam Indonesia, Yogyakarta Indonesia, who has given financial assistance for publication of this research.

IOP Conf. Series: Journal of Physics: Conf. Series **814** (2017) 012003 doi:10.1088/1742-6596/814/1/012003

Numerical simulations on the flow fields of dynamic axial compression columns in chromatography processes

Ru Chien Liang[1], Cheng Che Liu[2], Ming Tsai Liang[1] and Jiann Lin Chen[2,a]

[1] Dept. of Chemical Engineering, I-Shou University, Kaohsiung City 840, Taiwan
[2] Dept. of Mechanical and Automation Engineering, I-Shou University, Kaohsiung City 840, Taiwan

james88@isu.edu.tw

Abstract. Dynamic axial compression (DAC) columns are key elements in Simulated Moving Bed, which is a chromatography process in drug industry and chemical engineering. In this study, we apply the computational fluid dynamics (CFD) technique to analyze the flow fields in the DAC column and propose rules for distributor design based on mass conservation in fluid dynamics. Computer aided design (CAD) is used in constructing the numerical 3D modelling for the mesh system. The laminar flow fields with Darcy's law to model the porous zone are governed by the Navier-Stokes equations and employed to describe the porous flow fields. Experimental works have been conducted as the benchmark for us to choose feasible porous parameters for CFD. Besides, numerical treatments are elaborated to avoid calculation divergence resulting from large source terms. Results show that CFD combined with CAD is a good approach to investigate detailed flow fields in DAC columns and the design for distributors is straightforward.

1. Introduction

Simulated moving bed (SMB) chromatography process, which is a multicolumn chromatography process, has been applied to various industrial fields, such as foods, petrochemicals, fine chemicals, and pharmaceuticals [1]. The counter current movement of the adsorbent bed is simulated by periodically shifting the inlet and outlet nodes to the direction of the flow during each switching period. Therefore, the SMB process can achieve high productivity and low solvent consumption [2] and allows significantly higher yield and purity compared to a batch chromatography [3, 4]. Generally, the performance of chromatography depends on the combined effects of DAC column dynamics and the pre and post column geometry. In the beginning of the column, an even fluid distribution must be achieved cross sectionally through careful design in order to give satisfactory separation efficiency [5]. At the same time appropriate packing materials have to be chosen and the column has to be carefully packed to maintain the flow pattern as close as possible to a plug flow[6]. The above mentioned factors are especially important to columns with large diameter. If these factors were not considered, band distortion and unsymmetrical peaks would be observed because flow misdistribution can result from faulty header design [7, 8].
CFD has well been known for more than 20 years as a pragmatic and reliable tool to describe local hydrodynamics in various reactors and separation processes [9]. Its modelling has been applied to investigate the transport phenomena in the preparative chromatography column [10-12]. Therefore, based on the recent advances of flow characterisation by CFD, it is proposed to characterize in detail hydrodynamics of DAC columns, and then to deduce numerical treatments that can be beneficial to

simulations. We use CFD to access the internal information inside the DAC column and the flow is considered to be laminar due to low flow speed in the micro-particle packed column [13, 14]. In this study, we present the governing equations for the physical problem and introduce CAD and CFD techniques to simulate flow fields in the DAC column in section 2. Results and discussions are stated in section 3, which includes numerical treatments for enhancing calculation convergence and CFD code validations. Rules to design distributors at the header of the DAC columns, are also proposed in this section. Remarks are concluded in the final section, which includes the present contributions and future works.

2. Methodology

The present simulation is simplified as an incompressible laminar flow of a single species that does not generate heat. Inside the DAC column, no-slip boundary conditions are used for the walls and the packed bed is viewed as a porous medium in which the flow of a Newtonian fluid is known to follow Darcy's law at a low Reynolds number. The flowchart for the design of DAC columns in this study is illustrated in figure 1.

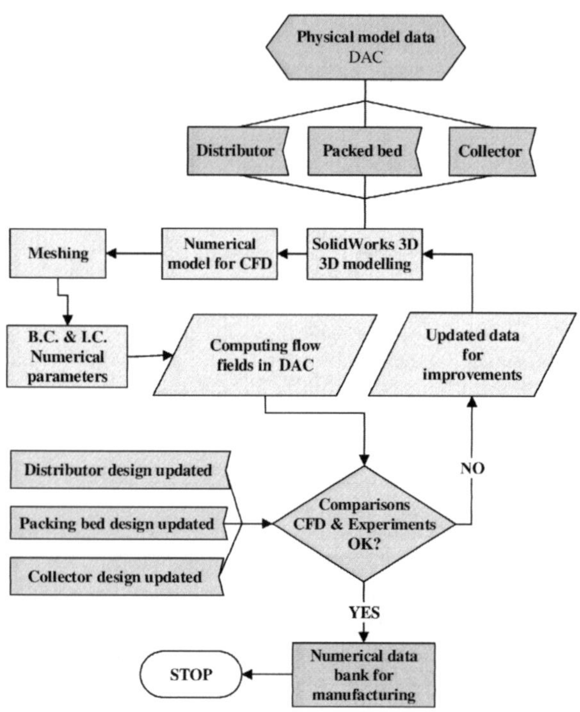

Figure 1. The flowchart for the design of DAC columns.

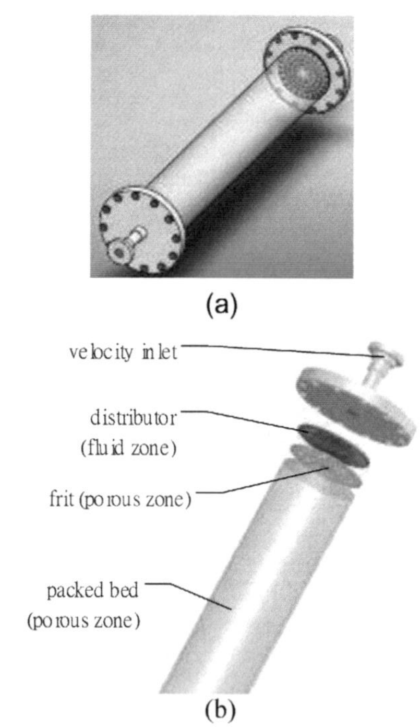

Figure 2. CAD diagrams of the DAC column (a) and components of its header part (b).

2.1. Governing equations

2.1.1. Equations in physical domain. The conservation equation for transport of a scalar quantity ϕ is demonstrated by the following equation written in integral form for an arbitrary control volume V as follows [15]:

$$\oint \rho\phi\vec{v}\cdot d\vec{A} = \oint \Gamma \nabla\phi \cdot d\vec{A} + \int_V S_\phi dV \quad (1)$$

where ρ indicates density; $\vec{v}$ is velocity vector; $\vec{A}$ is surface area vector; V is volume; Γ is the diffusion coefficient for ϕ and S_ϕ is source term of ϕ per unit volume. The symbol ϕ can be replaced by 1 for the continuity equation, and u, v and w are the momentum equations in the X, Y and Z directions, respectively. Inside the DAC column, the packed bed can be simulated by a porous medium, which is expressed as extra source terms including viscous loss (Darcy's law) and inertial loss. The added source terms to model the porous effect are expressed as

$$S_j = -\left(\frac{\mu}{\alpha} v_j + C' \frac{1}{2} \rho |v| v_j\right) \quad (2)$$

where α is permeability (1/α is viscous resistance); C' indicate inertial resistance factor and $|v|$ is the magnitude of velocity.

2.1.2. Equations in computational domain. Since the fluid used in this study is IPA (Isopropyl alcohol), the density is assumed to be constant in the flow field. Equation (1) can be expressed in generalized coordinates using the finite volume method approach for simulations. The discretized formulation for each cell (control volume) based on Eq. (1) yields

$$\sum_{f}^{N_f} v_f \phi_f A_f = \sum_{f}^{N_f} \Gamma_\phi (\nabla \phi_f)_n A_f + S_\phi V \quad (3)$$

where f denotes face of the control volume; N_f is number of faces enclosing cell.

2.2. CAD for 3D modelling

For most preparative and process scale chromatographic columns, DAC column ends include distributors (or collectors) and frits to distribute (or collect) fluid over the cross section of the column. Figure 2 displays CAD diagrams of the DAC column (2a) and the components of its header part (2b) by SoldWorks 3D CAD software. The 3D solid modellings from CAD are imported into the pre-processor in ANSYS MESHING for building the mesh system [16]. Figure 3 displays grid coordinates (left-hand side; LHS) and the grid distribution around column header (right-hand side; RHS) in the mesh system by the grid generation technique. The mesh is unstructured grid system with mixed hexahedra and tetrahedral grids. To effectively compute the present flow fields, the flow domain was split into smaller subdomains. We employed the mixed grid types to build the mesh around the distributor and collector. In the packed bed zone, structured hexahedral grids were used to fit the space inside the column. Moreover, finer grids were imposed along boundary walls to resolve the laminar boundary layer. The grid number for our parameter study was determined to be 0.8 million from grid independent tests and validated with experiments.

2.3. CFD for detailed flow fields

The equations governing the present incompressible flow fields have no link between the continuity and momentum equations due to lack of the equation of state. To solve the set of algebraic equations derived from the governing equations, the SIMPLE algorithm was employed to carry out the calculations [16, 17]. SIMPLE is an acronym for Semi-Implicit Method for Pressure Linked Equations. SIMPLE actually solves the pressure correction equation, which is derived from combining the discrete continuity and momentum equations. Therefore, the set of momentum and continuity equations are coupled and nonlinear so we solve the equations iteratively. The momentum equations are solved for the velocities; however, the newly obtained velocities don't satisfy continuity. Corrections to velocities and pressure are proposed to satisfy the discrete continuity equation in SIMPLE algorithm. After iterations for the above correction computations, the final solution can be obtained if the estimated residuals approach within required tolerances. In this study, the 3D numerical geometry model for the DAC column is created based from the engineering drawing as shown in figure 2. Then we use CFD pre-processor, ANYSYS MESHING to transform the geometry model to the mesh system as shown in figure 3. After the mesh system is imported to the CFD FLUENT solver,

IOP Conf. Series: Journal of Physics: Conf. Series **814** (2017) 012003 doi:10.1088/1742-6596/814/1/012003

numerical parameters, such as fluid properties, initial and boundary conditions have to be input. Besides, parameters for computation iterations and post processing for contour plotting need to be provided.

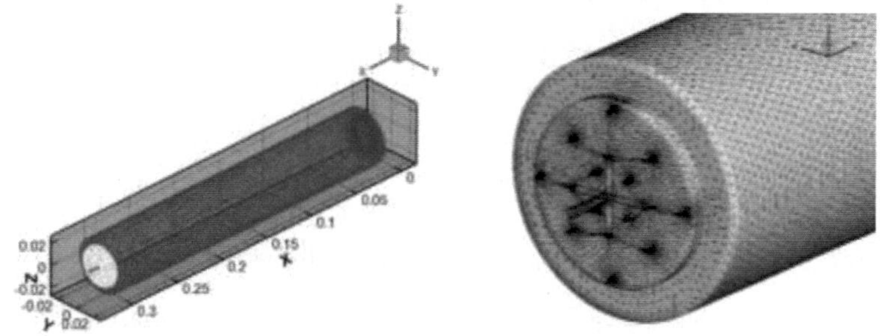

Figure 3. Grids in the mesh system: coordinates (LHS) and grids around the column header (RHS) by the grid generation.

3. Results and discussions

3.1. Numerical treatments

In this study, when the standard solution procedures and solution parameter settings were used, we found that the rate of convergence in computations slowed down if the pressure drop was relatively large in the flow direction in a porous region. This slow convergence can occur because the pressure drop of porous media appears as a momentum source term, which yields a loss of diagonal dominance in the matrix of discretized equations solved. The best remedy for poor convergence of a problem involving a porous medium is to supply a good initial guess for the pressure drop across the medium. Another possible way to deal with poor convergence in computation is to temporarily disable the porous media model and obtain an initial flow field without the effect of the porous region. With the porous media model turned off, FLUENT will treat the porous zone as a fluid zone and calculate the flow field accordingly. Once an initial solution is obtained, or the calculation is proceeding steadily to convergence, we thus enable the porous media model and continue the calculation with the porous region included. Based on the Kozeny-Carman correlation in Eq. (4) and following the solution strategies as mentioned above, we can obtain the convergent calculations as anticipated.

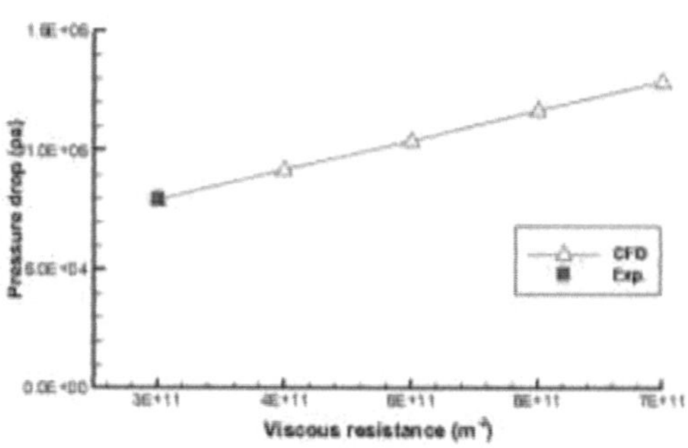

Figure 4. Seeking the appropriate viscous resistance via CFD.

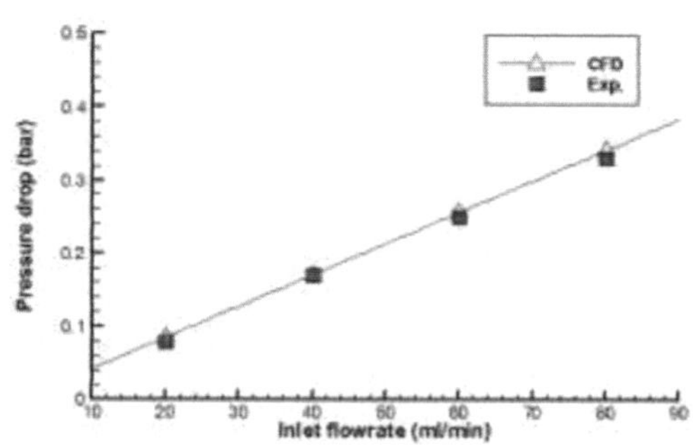

Figure 5. Pressure drop versus inlet flowrate for the DAC column.

The Kozeny-Carman correlation is represented by the following equation [14]:

$$\alpha = \frac{d_p^2 \cdot \varepsilon^2}{150(1-\varepsilon)^2} \tag{4}$$

where α is permeability; ε is porosity and d_p is diameter of packing particles. The porosity is the volume fraction of fluid within the porous region (i.e., the open volume fraction of the medium). The porosity is set to be 1.0 (or permeability $\alpha \to \infty$; viscous resistance $1/\alpha \to 0$) if the medium is treated as merely the IPA fluid. According to data from the manufacturer of the DAC column; $d_p = 50\times10^{-6}$m; $\varepsilon = 0.37$; $\alpha = 5.75\times10^{-12}\text{m}^2$. Therefore, the viscous resistance is evaluated to be $1.74\times10^{11}\text{m}^{-2}$. Thus at the beginning of calculations, this value is a trial to seek the appropriate one which meets the experimental data within the required tolerance as illustrated in figure 4. In figure 4, the experimental datum was searched out via the CFD by the trial value as an initial input from right-hand side to left in the horizontal axis. In fact, the process is a kind of reverse engineering to obtain the values we anticipate.

3.2. Code validation

The experiments were conducted in the Center for Advanced Chromatographic Processing at I-Shou University. Table 1 lists experiment parameters and these data are also applied for boundary conditions in CFD simulations. The working fluid is IPA, which dissolves a wide range of non-polar compounds, and is relatively non-toxic as compared to other solvents.

Table 1. Physical parameters for analysis (the working fluid is IPA)

Property	Value	Used in exp. or CFD
Viscosity	0.0023703kg/m-s	Exp. & CFD
Density	786kg/m^3	Exp. & CFD
Inlet velocity (exp.)	0.1062m/s; 0.2122m/s; 0.3183m/s; 0.4244m/s	Exp. & CFD
Porosity (DAC)	0.37	Exp. & CFD
Viscous resistance	Frit: 3.5×10^{11} m^{-2}; Packed bed: 3×10^{11} m^{-2}	CFD
Pressure	0∽200 bar	Exp. & CFD

According to our experiment setup, the velocity flowrate at the inlet can be adjusted by the air compressor, which supplies the loading pressure for the hydraulic cylinder to push downward the working fluid through the packed bed. The gauge pressure to provide the driving force in experiments can be reached to as high as 200 bar. Figure 5 illustrates the relationship of pressure drop versus inlet flowrate for the specific DAC column in experiments and CFD calculations. Experimental and CFD data agree well and the curve reflects their linear property. A linear function is proposed to correlate the CFD data for further applications. The correlation equation referring to pressure drops versus inlet flowrates for the present column can be expressed as

$$Y = 7.45\times10^{-9} + 4.26\times10^{-3} X \approx 4.26\times10^{-3} X \tag{5}$$

where Y denotes pressure drop (bar); X is inlet flowrate (ml/min). In fact, in porous media, pressure drop is typically proportional to velocity, which is described by the Darcy's law.

3.3. Rules for distributor design

A distributor (collector) plays a key role to make the plug flow in a DAC column. Manufacturers in this field have been devoted to improving the distributor's performance to raise the efficiency of their DAC column products [18, 19]. In this study, we have found the CFD is a good alternative way to design distributors in analytic approaches. Figure 6 shows the comparison of velocity vectors before the frit at the column header without a distributor (LHS) and with a distributor (RHS). Obviously, the distributor helps to forming plug flow at inlet and CFD helps to providing the visualization of vector fields to see the distributor's contribution. In addition, we have employed the CFD to investigate the flow features from two different distributors. Figure 7 displays an original distributor (LHS) we used and the modified one (RHS) we proposed. Axial velocity vectors for various locations in the column header from CFD have shown that the fluid speed is very low compared to the inlet velocity. Based on the same locations around the headers, we compare the average axial velocities (using log scale) from both distributors' CFD results for two inlet velocities (V1 and V2) as shown in figure 8. The analytic axial velocities were computed at the middle location of the DAC column based on the given inlet flowrate and the mass conservation law. It reveals that the average axial velocity of using the modified distributor approaches to the analytic value faster than using the original distributor for either V1 or V2. Obviously, the modified distributor performs better than the original one because of mass conservation considered in the design process.

The modified distributor was designed according to some rules proposed by this study. The rules are on the assumption that the plug flow is inviscid without circulation existing in channels of the distributor. The rules are made by satisfying the law of mass conservation and stated as follows:

1. Draw a radial line from the center of the distributor surface to get a radius and evenly divide the radius to N segments.
2. Make N-1 concentric circles on the surface to form N-1 rings and an inner circle in which the feed tube (inlet) is located. Calculate the areas and denote them S_1 to S_N from inner to outer region on the distributor surface.
3. Determine M_n (n=1 to N-1) uniformly spaced distribution openings surrounding each ring. Denote the opening area as A_n (No restrictions for the shape of the opening).
4. Make sure opening-area density, $(\sum_1^{M_n} A_n)/S_n$, on each ring (n) have the same value; it will keep the same total flowrate for each ring.
5. Denote the segment lengths of each channel (including its branch) from the inlet to the opening as Lj (j=1~J, J denotes total segments), then make sure total length $(\sum_1^{J} L_j)$ is the same for each channel on the surface. Besides, the cross-sectional area from the inlet to the opening along the channel keep invariant to make flowrate conservation.

Following the step of rules, the previous modified distributor and the example model with new pattern of opening distribution were proposed in figure 9. In figure 9, N is three; M_n is four; J is two with open-area density equal to 8.83%.

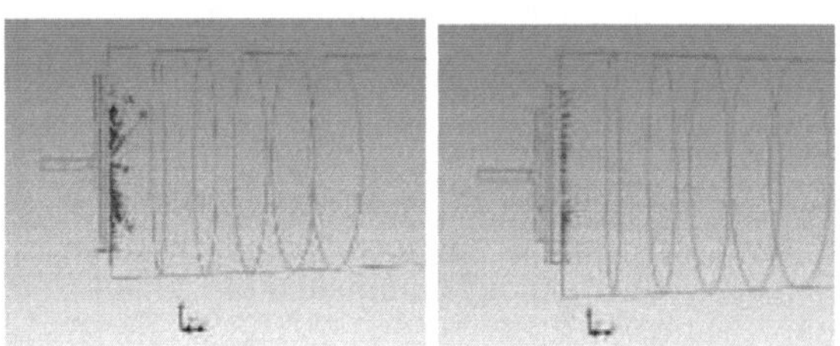

Figure 6. Velocity vectors without a distributor (LHS) and with a distributor (RHS) around headers.

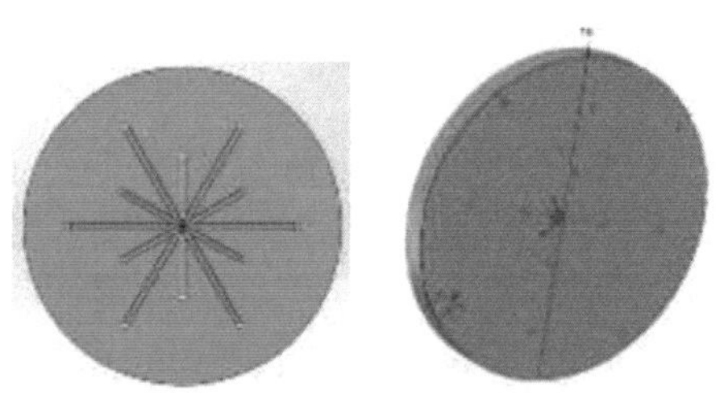

Figure 7. An original distributor (LHS) and the modified one (RHS).

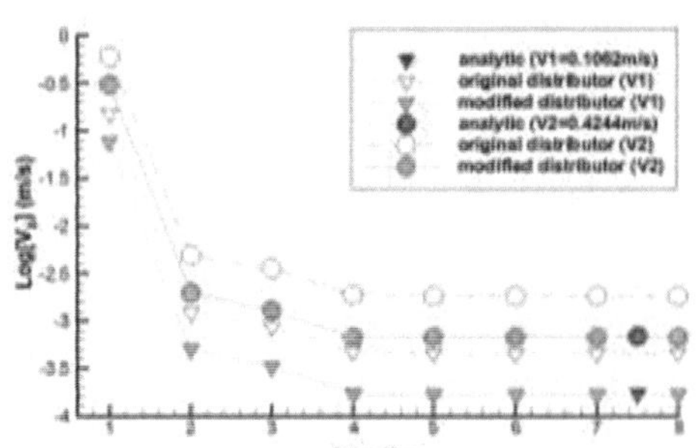

Figure 8. Average velocity distributions in the axial direction.

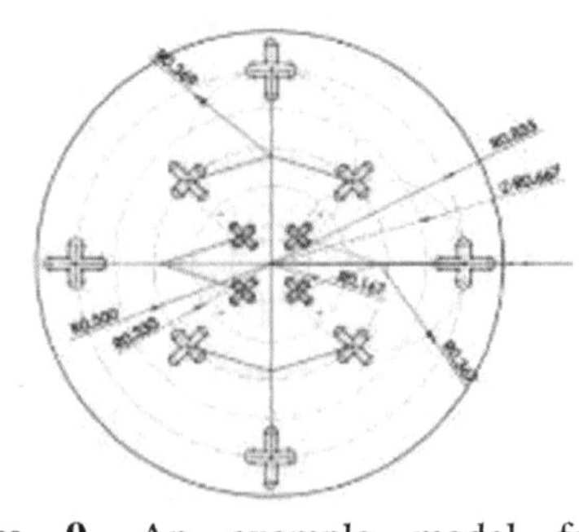

Figure 9. An example model for the distributor by proposed rules.

4. Concluding remarks

In this study, we have provided evidences by CFD simulations to show that the distributor plays a key role on velocity profile at inlet of the DAC column. Besides, the parameter of viscous resistance in packed bed can be numerically determined via the procedure of reverse engineering. Experimental and numerical data suggest the correlation equation for the relationship about pressure drop versus inlet flowrate for the present DAC column. Moreover, rules for designing the distributor have been proposed according to mass conservation law. New distributors thus can be designed based on the rules and an example model is presented. It is anticipated that based on the present study, transient cases involving concentration transport for columns with low length-to-diameter aspect ratios are worth further researches.

5. References

[1] Guiochon G 2002 *Journal of Chromatography A* **965** 129-61
[2] Song J Y, Kim K M and Lee C H 2016 *Journal of Chromatography A* **1471** 102-17
[3] Liang M T and Liang R C 2012 *Journal of Chromatography A* **1229** 107-12
[4] Liang M T, Liang R C, Lin C H, Hsu P J, Wu L Y, Chen H F, Wu Y W and Lee W C 2013 *The Journal of Supercritical Fluids* **79** 324-9
[5] Yuan Q S, Rosenfeld A, Root T W, Klingenberg D J and Lightfoot E N 1999 *Journal of Chromatography A* **831** 149-65
[6] Park J C, Raghavan K and Gibbs S J 2002 *Journal of Chromatography A* **945** 65-81
[7] Lode F G, Rosenfeld A, Yuan Q S, Root T W and Lightfoot E N 1998 *Journal of Chromatography A* **796** 3-14
[8] Wu Y and Ching C 2003 *Chromatographia* **57** 329-37
[9] Shyy W 2014 *Computational modeling for fluid flow and interfacial transport* (New York: Courier Corporation)
[10] Choopakdee W, Prasitchoke P, Shotipruk A and Ponpesh P 2014 *Computer Aided Chemical Engineering,* ed P S V Jiří Jaromír Klemeš and L Peng Yen (New York: Elsevier) pp 1393-8
[11] Fangueiro Gomes L, Augier F, Leinekugel-le-Cocq D, Vinkovic I and Simoëns S 2015 *Chemical Engineering Science* **132** 46-58
[12] Fangueiro Gomes L, Augier F, Leinekugel-le-Cocq D, Vinkovic I and Simoëns S 2016 *Chemical Engineering Science* **153** 188-98
[13] Cohen Y and Metzner A 1981 *AIChE Journal* **27** 705-15
[14] Li Y and Park C-W 1998 *Industrial & engineering chemistry research* **37** 2005-11
[15] Jeng Y N and Chen J L 1992 *Numerical Heat Transfer, Part B: Fundamentals* **22** 211-34
[16] *ANSYS FLUENT 13 Theory Guide* 2010 (USA: ANSYS, Inc.)
[17] Patankar S V 1980 *Numerical heat transfer and fluid flow* (New York: Hemisphere Pub. Co)
[18] Allen C C and Durham N C 1985 United States Patent, US 4537217
[19] Gebauer K, Kranse J and Lundkvist M 2012 United States Patent, US 2012/0074051

Acknowledgments

Financial Assistance for this research provided by the Taiwanese Ministry of Science and Technology, under contract number MOST 105-2218-E-214-002, is greatly appreciated.

Numerical study of MPS method with large eddy simulation for fluid solid coupling problem

Chao YANG, Huaixin ZHANG, Huilan YAO

School of Naval Architecture, Ocean and Civil Engineering;
Shanghai Jiaotong University, Shanghai 200240, China

hxzhang@sjtu.edu.cn

Abstract. The Moving-Particle Semi-implicit method (MPS) is a kind of meshless Lagrangian calculation method. This method uses particles instead of mesh. In the pretreatment it works simply and conveniently and has high computational efficiency. In practical engineering, many of fluid problems are turbulent flows. Large eddy simulation is a major means of studying turbulence. Fluid-structure coupling is an independent branch of mechanics combined with fluid dynamics and solid mechanics, which is the hot and difficult area of research in many fields at present. In this paper, for the numerical simulation of turbulent flow with interaction of fluid-structure, the modified MPS-LES method is applied in two dimensional dam-break problem. It proves that MPS-LES method can be extended on solving the fluid-solid coupled problem.

1. Introduction

Fluid-structure coupling problem is common in the field of ship engineering and marine technology. In computational structural mechanics, the most widely used numerical discretization method is finite element method. But in computational fluid dynamics, it primarily adopts the finite difference method or finite volume method. Therefore, it not only needs to take both fluid-structure interactions in numerical analysis but also need in the coupling between the methods into account. In the process of fluid-structure coupling analysis, Discontinuous interface between the fluid and solid grid will be produced. In dealing with such time-varying discontinuous interface, the grid method mainly restructures the grid in every calculated step so as to guarantee the consistency of grid lines with the interface after the deformation. The main strategy calculation method based on grid is used that reconstruction in each calculation step in the grid, to ensure that the grid line is still consistent with the deformation of the interface. It will not only face the numerical difficulties, but also decreases accuracy. Sometimes too much mesh deformation will result in distortion and grid computing terminates.

MPS is a new Lagrangian particle method, which was first proposed by Koshizuka, S, Oka, Y.[1] in 1996. Due to a very extensive applications of MPS, after it was proposed[1],[2] , it has been widely used in many research fields. In 1999, Gotoh[3] used the MPS method to simulate wave propagation. In practical engineering problems, a lot of flowing fluid is turbulent flow, and large eddy simulation is a major means of research on turbulence. With respect to the ordinary governing equation, subgrid Reynolds stress terms are added in Large eddy simulation (LES), so large eddy simulation method is very suitable for MPS. In 2001 Gotoh[4], for the first time, applied large-eddy turbulence model to MPS method. In 2001, Chikazawa[5] adopted MPS method in the fluid-structure interaction, and It is the first time that this method is applied in computational structural mechanics. In 2008 Pan Xu Jie [6][7][8][9] has been carried out research on the main problems existing in MPS method. He developed a new technique of free surface recognition. In 2013 based on MPS-LES method ,Yu Qian[10] simulated the impact of a wedge into the water problem. In the ocean wave environment, fluid-structure coupling numerical simulation has great significance on the interaction between ocean wave and marine construction. During the early development of the MPS method, many studies have mainly focused on the simulation of fluid. But research on fluid-structure coupling problems are very rare. This paper has initiatively applied MPS method with Large Eddy Simulation into the elastic fluid-structure coupling calculation. The compared and analyzed simulation results show that MPS-LES method can be well applied in the solution of fluid-structure coupling problem.

2. MPS method

Adopted to deal with the fluid-structure coupling problem, First, ignore the properties of the solid particles, they're treated as fluid particles, the MPS method can also be adopted for all particles. When using large eddy simulation method to simulate the turbulent flow, the fluid control equation with equation should be added sub-grid Reynolds stressing force item. The calculation method for fluid is the same as our published paper, detail see Ref[9]. The speed and position of solid particles is modified in accordance with elastomer control equation.
The control equations of elastomer is as follows:

$$\rho\frac{Dv_\alpha}{Dt}=-\frac{\partial p}{\partial x_\alpha}+\frac{\partial\sigma_{\alpha\beta}}{\partial x_\beta} \tag{1}$$

where ρ is the material density, p is an isotropic pressure which is got in particle location , $\sigma^{\alpha\beta}$ is unisotropic components of stress tensor .

$$\sigma_{ij}=2\mu\varepsilon_{ij}^{n}=\frac{2\mu(u_{ij}\cdot r_{ij})r_{ij}}{\left|r_{ij}^{0}\right|\left|r_{ij}\right|^{2}} \tag{2}$$

$$\tau_{ij}=2\mu\varepsilon_{ij}^{s}=2\mu\frac{u_{ij}}{\left|r_{ij}^{0}\right|}-\sigma_{ij} \tag{3}$$

$$p=-\lambda(\varepsilon_{rr})_i=-\lambda\frac{d}{n^0}\sum_{j\neq i}\frac{u_{ij}\cdot r_{ij}}{\left|r_{ij}^{0}\right|\left|r_{ij}\right|}w(\left|r_{ij}^{0}\right|) \tag{4}$$

σ_{ij} and τ_{ij} are the Normal stress and shear stress to the particle i and j, and u_{ij} is the variation of the relative position between the particles, r_{ij} is the relative position between particle i and j, d is the number of space dimension.
The acceleration produced by the force between particles is:

$$\left[\frac{\partial v_i}{\partial t}\right]_n=\frac{2d}{\rho_i n^0}\sum_{j\neq i}\frac{\sigma_{ij}}{\left|r_{ij}^{0}\right|}w\left(\left|r_{ij}^{0}\right|\right) \tag{5}$$

$$\left[\frac{\partial v_i}{\partial t}\right]_s=\frac{2d}{\rho_i n^0}\sum_{j\neq i}\frac{\tau_{ij}}{\left|r_{ij}^{0}\right|}w\left(\left|r_{ij}^{0}\right|\right) \tag{6}$$

$$\left[\frac{\partial v_i}{\partial t}\right]_p=-\frac{d}{\rho_i n^0}\sum_{j\neq i}\frac{(p_i+p_j)r_{ij}}{\left|r_{ij}^{0}\right|\left|r_{ij}\right|}w\left(\left|r_{ij}^{0}\right|\right) \tag{7}$$

$\left[\frac{\partial v_i}{\partial t}\right]_s$ is the shear acceleration , $\left[\frac{\partial v_i}{\partial t}\right]_n$ is normal acceleration, $\left[\frac{\partial v_i}{\partial t}\right]_p$ is the acceleration produced by the pressure.

Updating the particles' rotation angle speed, rotation angle, speed, and coordinate position, could be described as follows:

$$v_i^{n+1} = v_i^n + \Delta t \left[\frac{\partial v_i}{\partial t} \right]^n \tag{8}$$

$$r_i^{n+1} = r_i^n + \Delta t v_i^{n+1} \tag{9}$$

3. Results of numerical simulation

Dam-break model is shown below, the initial water column length and width were 0.292m, 0.146m, height of 0.08m, width of 0.012m, flat located at x = 0.292m. Set flat elastic modulus of 1.0*pa, Poisson's ratio of 0.0, a density of 2.5 times than that of water. Take the dame break model to particle model .The results shown in figure 1, the particle spacing $l_0 = 0.0025m$, $r_e = 2.1l_0$, time step takes 0.0001s. In this simulation, three models have been selected, they're ideal fluid, the viscous laminar flow and LES(large eddy simulation) based on Smagorinsky eddy viscous model respectively.

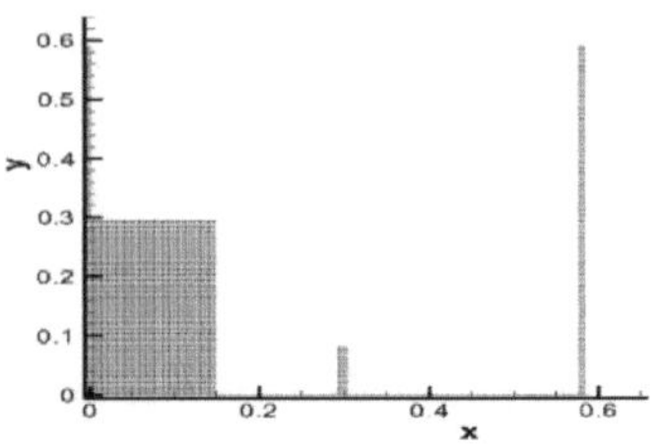

Figure 1.System of dam break example with elastic structure

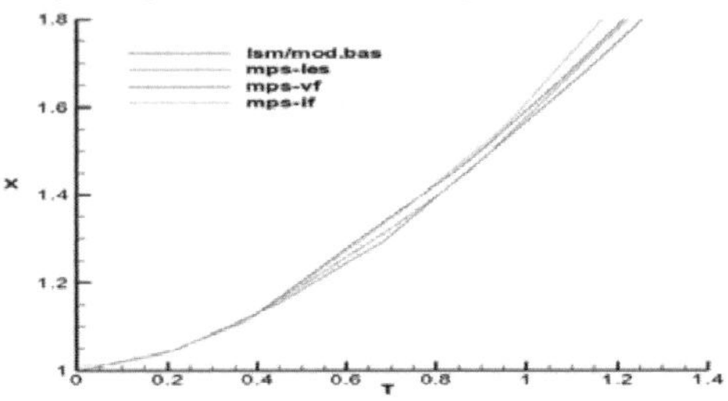

Figure 2. Leading edges of the water column collapse

See figure 2 , no matter what the conditions is selected in simulation, the water column quickly collapsed under the force of gravity, but in different simulated conditions, we can see a slight difference in the speed of the collapse, in which the ideal fluid was first to reach flat position. Because the ideal fluid does not consider any sticky, it collapse faster. The large eddy simulation in addition to viscosity, takes eddy viscosity into account, which is the largest internal energy consumption, so it arrives at the flat position latest. Time and distance in the figure is non-dimensional (X=x/l,T=t(2g/l)1/2).The results are in agreement with reference [11](shown as ism/mod.bas in fig.2)
Large Eddy Simulation of dam-break fluid-structure coupling calculation results are shown in figure 3, it can be seen from the figure that when applied on coupled fluid-structure calculations, The improved MPS method based on large eddy simulation can accurately simulate the large deformation of fluid free surface accompanied with the breaking and splashing, and elastic deformation of elastic plate under the impact of fluid.

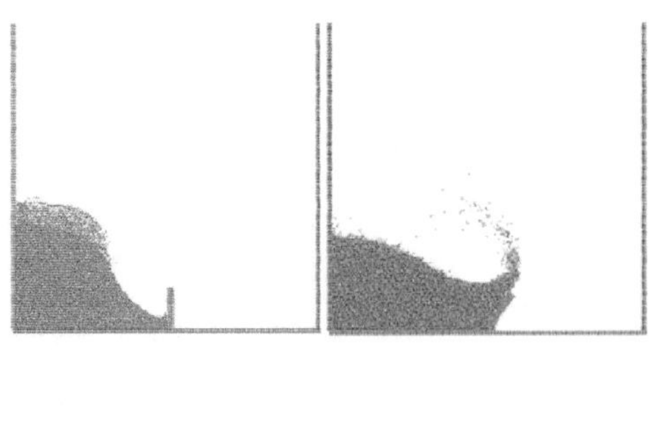

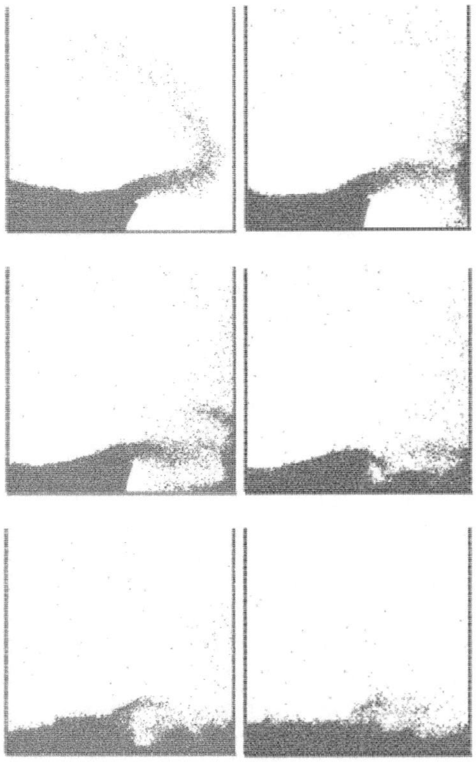

Figure 3. Interaction between dam-break impact elastic plate.

Under the impact of fluid the plate at the top will produce deformation, and figure 4 shows the variation of flat plate at the top displacement with time. The top of flat deform by the impact of water compared with other numerical results [12] and is illustrated in figure 4. Under the impact load of flow, the displacement and direction of the top of the plate vary with the time, reciprocating oscillating movement occurs at the top of flat plate.

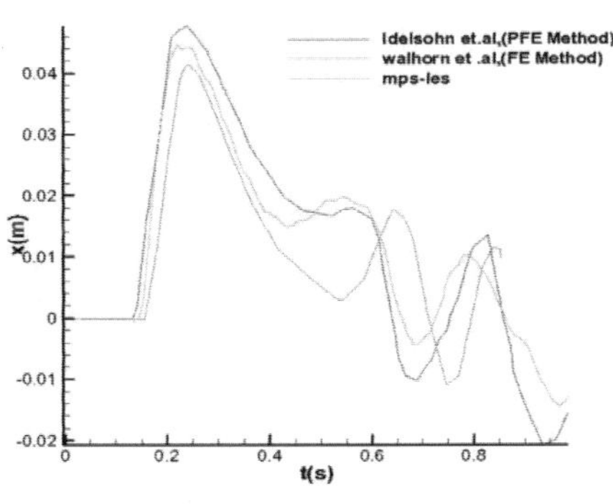

Figure 4. History of the displacement

4. Conclusion

In this paper, the fluid-structure interaction of dam-break model is simulated by large eddy simulation moving particle semi-implicit method, in which the calculation of attacking pressure were improved and oscillation were reduced. And the position of head of the fluid, and the elastic deformation and displacement of the plate were analyzed. The MPS method based on large eddy simulation is applied on the dam-break fluid-solid coupling calculation, accurately simulates large deformation of the fluid free surface even with the breaking and splashing situation, and the deformation of elastic plate under the impact of the fluid. It confirmed that the moving particle semi-implicit method combined with large-eddy simulation applying to fluid-structure interaction problems is reasonable and correct.

5. References

[1] Koshizuka, S. and Oka, Y. Moving-Particle Semi-Implicit Method for Fragmentation of Incompressible Fluid, Nuclear Science and Engineering, 1996, 123: 421~434.

[2] Koshizuka, S., Tamako, H. and Oka, Y. A Particle Method for Incompressible Viscous Flow with Fluid Fragmentation, Journal of Computational Fluid Dynamics, 1995, 4: 29~46.

[3] Gotoh, H. and Sakai, T. Largrangian Simulation of Breaking Waves Using Particle Method, Coast Engineering Journal, 1999, 41: 303~326.

[4] Gotoh, H., Shibahara, T. and Sakai, T. 2001. Sub-particle-scale turbulence model for the MPS method - Lagrangian flow model for hydraulic engineering, Computational Fluid Dynamic Journal, 9(4): 339~347.

[5] Chikazawa,Y., Koshizuka, S. and Oka, A Particle Method for Elastic and Visco-Plastic Structures and Fluid-Structure Interactions, Computational Mechanics. 2001, 27: 97~106.

[6] Pan, X.J., Zhang, H.X. and Lu, Y.T. Moving-Particle Semi-Implicit Method for Vortex Patterns and Roll Damping of 2D Ship Sections, China Ocean Engineering, 2008, 22:399~407.

[7] Pan, X.J., Zhang, H.X. A study on pressure fluctuation in sloshing simulation of moving-particle semi-implicit method, Journal of hydrodynamic, 2008, 23(4):453~463. (in Chinese)

[8] Pan, X.J., Zhang, H.X. Moving-Partical Semi-Implicit Method for Simulation of Liquid Sloshing on Roll Motion, Journal of Shanghai Jiaotong university, 2008, 42(11):1904~1907. (in Chinese)

[9] Pan, X.J. , Zhang, H.X, Sun X.Y. Numerical simulation of sloshing with large deforming free surface by MPS-LES method, China Ocean Engineering, 2012 Dec, Vol.26(4), pp.653-668

[10] YU Qian , Zhang, H.X.. Research on water entry of wedge based on the improved MPS method with large eddy simulation, The ocean engineering, 2013,9-15,90 (in Chinese)

[11] Walhorn E, Kolke A, Hubner B, Dinkler D.L: Fluid - structure coupling within a monolithic model involving free surface flows, Computers and Structures 83 (2005) 2100 - 2111

[12] Ashkan R P. An SPH projection method for simulating fluid-hypoelastic structure interaction, Comput. Methods Appl. Mech. Engrg. 198 (2009) 2785 - 2795

Acknowledgments

This work is supported by the National Natural Science Foundation of China (No. 51479116 and No. 11272213).

IOP Conf. Series: Journal of Physics: Conf. Series **814** (2017) 012005 doi:10.1088/1742-6596/814/1/012005

First order transition in two dimensional coulomb glass

Preeti Bhandari[1] and Vikas Malik[2]

1Department of Physics, Jamia Millia Islamia, New Delhi 110025, India
2Department of Physics and Material Science, Jaypee Institute of Information Technology, Uttar Pradesh, 201304, India

preetibhandari15@gmail.com

Abstract. We have studied the ground states of two dimensional lattice model of Coulomb Glass via Monte Carlo annealing. Our results show a possibility of existence of a critical disorder (W_c) below which the system is in the charge ordered phase and above it the system is in the disordered phase. We have used finite size scaling to calculate $W_c = 0.2413$, the critical exponent of magnetization $\beta = 0$ indicating discontinuity in magnetization and the critical exponent of correlation length $\nu = 1.0$. The distribution of staggered magnetization for different disorder strengths shows a three peak structure. We thus predict that two dimensional Coulomb Glass shows a first order transition at T=0.

1. Introduction

Since the introduction of Random field Ising Model, a lot of discussion has been done on the possibility of existence of ferromagnetic ordering below a critical disorder strength (W_c) in two dimensional (2D) RFIM. From Imry-Ma arguments [1], one finds that the lower critical dimension (d_c) below which the ferromagnetic ordered phase gets destroyed is $d_c \leq 2$, where d = 2 was considered as a limiting case. By doing field theoretical calculations [2,3], later it was suggested that $d_c = 3$. In 1983, Binder [4] gave an argument that the roughening of domain walls would lead to the destruction of ferromagnetic ordering in 2d RFIM. Then an exact result was given by Bricmont and Kupiainen [5] in 1987 where they showed that there exist a ferromagnetic ordering in three dimensional (3d) RFIM. A rigorous theoretical proof was then given by Aizenman and Wehr [6] in 1989 that no long range ordering is possible in 2d RFIM case. Scaling theory for 3d RFIM assuming second order transition at T=0 leads to modified hyperscaling laws [7].

However, in contradiction to all the above arguments Frontera and Vives [8] in 1999 showed numerical signs of transition in 2d RFIM at zero temperature below a critical random field strength. In 2012, Aizenmann [9] too claims existence of phase transition in 2d RFIM. Spasojevic et al [10] have also given some numerical evidences which proves the existence of phase transition in 2d non-equilibrium RFIM at T = 0. Recently, Suman and Mandal [11] have studied some aspects of non-equilibrium 2d RFIM at low temperature. They have commented that 2d RFIM exhibits a phase transition in the disorder parameter even at T > 0. These numerical work suggests a possibility of presence of long range ordering in 2d RFIM at low disorder strengths.

If we assume that the above arguments which are claiming that at T = 0 there is a finite disorder strength below which long range ordering in 2d RFIM exist, then the order of transition is another question which is not answered very clearly. This question remains unclear even for 3d RFIM where there is no controversy over the presence of phase transition at zero temperature. Some earlier work suggested a first order transition [12-18] but there are arguments [19-22] which favour second order

transition in 3d RFIM. Since the value of magnetic component β/ν is very small, it is difficult to determine whether at transition the magnetization vanishes continuously or discontinuously. Numerical study of 3d RFIM at T = 0 claims transitions to be of continuous type [23-26] but the value of the critical exponent of magnetization (β) is very close to zero in all the papers.
We are here interested in determining whether a critical behaviour similar to 2d RFIM exist in the case of 2d Coulomb Glass (CG). Coulomb Glass is defined as a system where all electron states are localized and interactions takes place via long-range Coulomb potential. Since at low temperature, localized electrons are unable to screen the Coulomb interactions effectively, it will be interesting to see how differently these long range interactions effect the phase transition from the RFIM. In our previous work [27] we found the transition from COP to disordered phase was driven by rearrangement of domain wall of the metastable state in charge ordered phase (COP) as W was increased to give disordered phase which indicates phase coexistence around transition. In this paper we have calculated the critical exponents using finite size scaling analysis. We have also investigated the distribution of staggered magnetization which shows a three peak structure characterstic of a first order transition.

2. Model

We have considered lattice model of CG in 2d as defined by Davies et al [28]. The model is a regular lattice of sites where on each site electron states are localized. The Hamiltonian of this system, defined over a spin configuration $\{S_i = \pm 0.5\}$ is given as

$$H = \sum_i \phi_i S_i + \frac{1}{2}\sum_{i \neq j} J_{ij} S_i S_j \qquad (1)$$

where the random on-site energies φ_i are chosen independently from a probability distribution $P(\varphi_i)$ which is considered as

$$P(\phi) = \begin{cases} \frac{1}{2W}, & if\ -W/2 \leq \phi \leq W/2 \\ 0, & otherwise \end{cases} \qquad (2)$$

The unscreened Coulomb interactions is defined as $J_{ij} = e^2/\kappa R_{ij}$ where κ is the dielectric constant and R_{ij} is the distance between sites i and j which is calculated using periodic boundary conditions. Since the system under consideration possesses particle-hole symmetry hence the chemical potential μ = 0. We have measured energies in units of $e^2/\kappa a$ where a ≡ 1 is the lattice constant. We have only considered the case of half filling where the number of electrons are half of the total number of sites (N) in the system.
We assume that there exist a critical disorder W_c below which phase is charge ordered and above which it is disordered [29]. Now for a CG system a COP implies Antiferromagnetic ordering as $J_{ij} > 0$. So staggered magnetization is the order parameter which is defined as

$$M_s = \left\langle \frac{1}{N}\sum_i \sigma_i \right\rangle \qquad (3)$$

where $\sigma_i = (-1)^i S_i$ and <...> is the ensemble average performed over different disorder configurations.

3. Numerical simulation

We have first used Monte Carlo (MC) annealing technique to reach the minimum possible energy state. The initial configuration used for the simulation was completely random with half sites assigned S_i = 0.5 and the second half assigned with S_i = -0.5 . The annealing was done using the Metropolis algorithm [30]. This algorithm constitutes of a random walk in the entire space of all possible configurations of the system. Since the number of electrons are conserved, one uses Kawasaki Dynamics. A single MC step is performed by randomly choosing two sites i and j, where one site is unoccupied (S_i = 0.5) and the second site is occupied (S_i = -0.5) for spin-exchange. This exchange is not always executed, which is possible in a simple MC procedure. If the exchange of spins corresponds to relaxation i.e. if the final state of the system yields an energy level which is lower than the initial state, then the exchange is done with a probability

$$P_{ij} = 1 \tag{4}$$

On the other hand if the exchange corresponds to a thermal excitation into a state whose energy is higher than the initial state, let us assume by an amount ΔE_{ij} then the spin-exchange is done with a probability given as

$$P_{ij} = \exp\left(\frac{-\Delta E_{ij}}{\kappa T}\right) \tag{5}$$

where $\Delta E_{ij} = \varepsilon_j - \varepsilon_i - 1/R_{ij}$ is calculated using the single particle Hartree energy (ε_i) [31] which is expressed as

$$\varepsilon_i = \phi_i + \sum_{i \neq j} J_{ij} S_j \tag{6}$$

The Hartree energy of the initial random system is calculated and stored. If the chosen site is the original position of the electron then the Hartree energy remains unchanged but if the electron hops to a new site, then we update all the Hartree energies of the system. When the system is in the low temperature state then the electron chooses its initial state site with a very high probability, so we do not have to recompute our Hartree energies again which speeds up our simulation considerably at low temperature regime. Annealing was done from $\beta = 1$ to $\beta = 100$ (where $\beta = 1/T$) for different system sizes (L=16,32,48). At low temperature we increased the MC steps and our longest run was $5\text{x}10^5$ for each β above 20. The investigation were carried out at different disorders starting with W=0.0 to W=0.50 for all L.

Once annealing was over we then identified the clusters which were created in the minimum energy state using the Hoshen-Kopelman algorithm [32]. The cluster with nearest-neighbour sites that have same σ_i was defined as domain. After identifying the domains we then investigated each domain's interactions with all the other domains present in a single disorder realization. We did not find any significant domain-domain interactions present. The domains were then flipped one by one if it decreased the energy of the system. The final state in which no more flipping was possible was then considered as the ground state. By this method we do not claim that we obtained the exact ground state.

4. Results and discussion

After obtaining ground state, we investigated the average size of the largest and second largest domain at each disorder for the largest system size that we have considered (L=48). Figure 1 shows the volume of the largest and the second largest domains in the ground state. For W<0.20 all the domains got flipped and we found single domain picture which indicates a COP. But on further increasing the disorder strength the size of the second largest domains starts increasing indicative of occurrence of

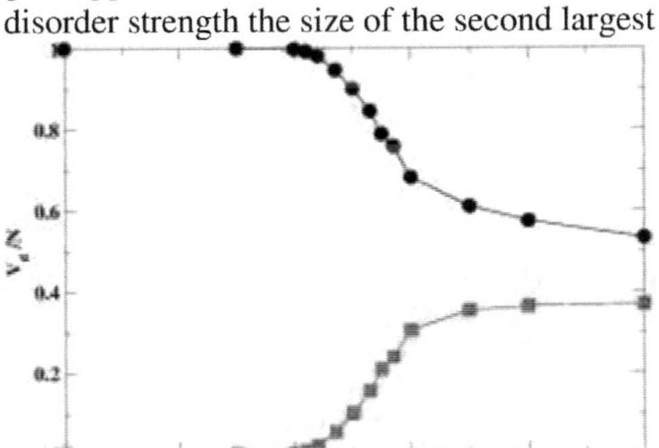

Figure 1. Largest (•) and second largest (■) domain size (V_d) divided by the system size $N = L^2$ at different disorder strength (W) for L=48

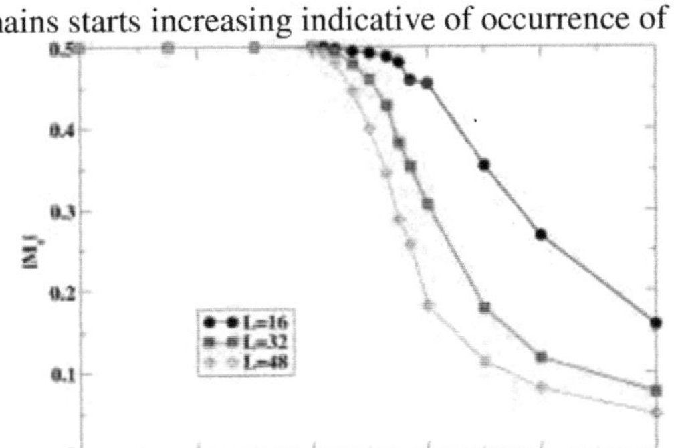

Figure 2. Behavior of staggered magnetization vs disorder strength W for system size L=16, 32, 48.

IOP Conf. Series: Journal of Physics: Conf. Series **814** (2017) 012005 doi:10.1088/1742-6596/814/1/012005

disordered phase. The picture is similar to 2d and 3d RFIM where the ordered state breaks into two large domains [33]. For each disorder strength (W) we have then recorded the average staggered magnetization per spin. The staggered magnetization shows sharp decrease with disorder as shown in figure 2. Calculation was done for different system sizes, L=16, 32, 48 and disorder averaging was done over 800, 400 and 350 realizations of random field respectively.

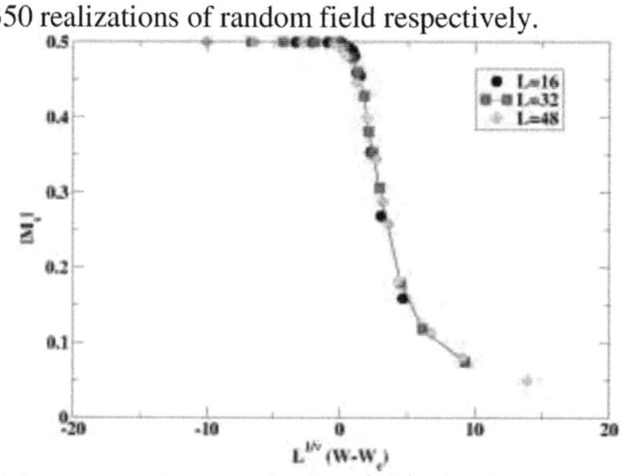

Figure 3. Scaling plot of the staggered magnetization $|M_s|$ with the parameters for $W_c = 0.2413$ and $1/\nu = 1.0$. The full line connects the points for L=32 as a guide for the eyes. Here the data for $|M_s|$ are not rescaled by the factor $L^{-\beta/\nu}$, which implies that $\beta = 0$.

To find the exact value of critical disorder (W_c) and to determine the order of transition we have used standard finite size scaling function for M_s which is defined as

$$M_s = L^{-\beta/\nu}\tilde{M}_s((W-W_c)L^{1/\nu}) \qquad (7)$$

where β is the exponent of staggered magnetization and ν is defined as correlation length exponent. The scaled data is shown in figure 3, where $W_c = 0.2413$; $\beta = 0$; and $1/\nu = 1.0$. The value of $\beta = 0$ could be indicative of a first order transition but as mentioned in the introduction is not a conclusive proof for a first order transition.

In our previous paper [27] we studied the system near the transition region where we found that the domain wall of the metastable state in the charge ordered phase (COP) shifts to give disordered ground state. This indicates that the transition is first order type as the free energy which is equal to the energy at zero temperature has three minimas which are centered around $M_s = \pm 0.5, 0$. This also shows that for each disorder configuration the staggered magnetization changes discontinuously at W_c. Such discontinous jumps for bond energy were also reported for 3d RFIM [18, 22, 34]. The domains of the metastable state in the COP and the domains in ground state of disordered phase are plotted in figure 4 for L=48.

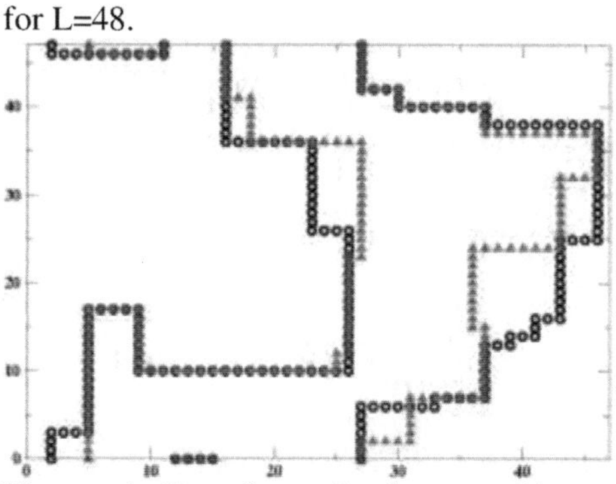

Figure 4. Domain wall of metastable state at W=0.235 and domain wall of ground state at W=0.25 are plotted for one of the configurations at L=48.

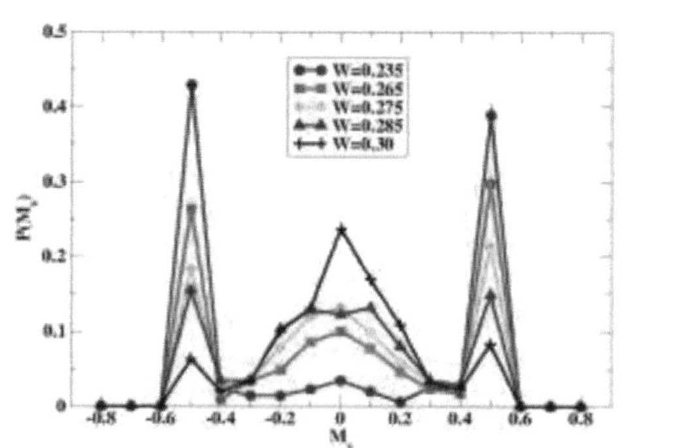

Figure 5. Histogram of the staggered magnetization distribution for L=48 system for different W's.

We have then investigated the distribution of statggered magnetization at different W for L=48. As shown in figure 5, for $W<W_c$, peaks are centered around $M_s = \pm 0.5$ and for $W>W_c$ around $M_s = 0$ as expected. In the transition region $0.265 < W < 0.285$, one gets a three peak structure. The transition region is same as one gets in the staggered magnetization vs W graph in figure 2 for L=48. This three peak picture around the transition region is a further indication of coexisting phases which is a characteristic feature of first order transition. A similar picture for 4d bimodal RFIM was reported by Swift et al [35] but no coexisting phases were found for 3d bimodal and Gaussian RFIM.

5. Conclusions

To summarize we have investigated the ground state of 2d lattice model of Coulomb Glass. Using finite size scaling analysis, we have shown existence of critical disorder below which the phase is ordered (antiferromagnetic ordering) and above $W_c = 0.2413$ it is disordered. The critical exponent of magnetization, $\beta = 0$ indicates that the transition is first order. To verify this point further we have also shown the presence of coexisting phases in two ways. Firstly, the domain in the metastable state of the COP shifts to give ground state in the disordered phase and secondly the three peak picture in the distribution of M_s around the transition region characteristic feature of first order transition.

References

[1] Imry Y and Ma S K 1975 *Phys. Rev. Lett.* **35** 1399
[2] Young A P 1977 *J. Phys. C* **10** L257
[3] Parisi G and Sourlas N 1979 *Phys. Rev. Lett.* **43** 744
[4] Binder K 1983 *Z Phys. B* **50** 343
[5] Bricmont J and Kupiainen A 1987 *Phys. Rev. Lett.* **59** 1829
[6] Aizenman M and Wehr J 1989 *Phys. Rev. Lett.* **62** 2503
[7] Bray A J and Moore M A 1985 *J. Phys. C* **18** L927
[8] Frontera C and Vives E 1999 *Phys. Rev. E* **59** R1295
[9] http://ricerca.mat.uniroma3.it/ipparco/convegno70/seminars.html
[10] Spasojevic D S Janicevic and Knezevic M 2011 *Phys. Rev. Lett.* **106** 175701
[11] Sinha S and Mandal P K 2013 *Phys. Rev. E* **87** 022121
[12] Houghton A, Khurana A and Seco F J 1985 *Phys. Rev. Lett.* **55,** 856
[13] Young A P and Nauenberg M 1985 *Phys. Rev. Lett.* **54** 2429
[14] Brezin E and De Dominicis C 1998 *Europhys. Lett.* **44** 13
[15] Machta J, Newman M E J and Chayes L B 2000 *Phys. Rev. E* **62,** 8782
[16] Fytas N G and Malakis A 2008 *Eur. Phys. J. B* **61** 111
[17] Fytas N G and M-Mayor V 2013 *Phys. Rev. Lett* **111** 019903
[18] Wu Y and Machta J 2006 *Phys. Rev. B* **74** 064418
[19] Villian J 1985 *Phys. (Paris)* **46** 1843
[20] Fisher D S 1986 *Phys. Rev. Lett.* **56** 416
[21] Natterman T 1997 *Spin Glasses and Random Fields*, ed Young A P (Singapore: World Scientific)
[22] Hartmann A K and Young A P 2001 *Phys. Rev. B* **64** 214419
[23] Middleton A A and Fisher D S 2002 Phys. Rev. B **65** 134411
[24] Dukovski I and Machta J 2003 *Phys. Rev. B* **67** 014413
[25] Ogielski 1986 *Phys. Rev. Lett.* **57**, 1251
[26] Vink R L C, Fischer T and Binder K 2010 *Phys. Rev E* **82** 051134
[27] Bhandari P and Malik V 2016 *arxiv:1610.06750v1* [cond-mat.dis-nn]
[28] Davies J H, Lee P A and Rice T M 1984 *Phys. Rev. B* **29** 4260
[29] Goethe M and Palassini M 2009 *Phys. Rev. Lett.* **103** 045702
[30] Metropolish N, Rosenbluth A W, Rosenbluth M N, Teller A H and Teller E 1953 *J. Chem. Phys.* **21** 1087
[31] Efros A L and Shklovskii B I 1975 *J. Phys. C* **8**, L49
[32] Hoshen J and Kopelman R 1976 *Phys. Rev. B* **14,** 3428
[33] Esser J, Nowak U and Usadel K D 1997 *Phys. Rev. B* **55** 5866

IOP Conf. Series: Journal of Physics: Conf. Series **814** (2017) 012005 doi:10.1088/1742-6596/814/1/012005

[34] Wu Y and Machta J 2005 *Phys. Rev. Lett* **95** 137208
[35] Swift M R, Bray A J, Maritan A,Cieplak M and Banavar J R 1997 *Europhys. Lett.* **38** 273

Acknowledgments
We wish to thank NMEICT cloud service provided by BAADAL team, cloud computing platform, IIT Delhi for the computational facility. Preeti Bhandari acknowledges UGC, Govt. of India for financial support through UGC-BSR fellowship (F.25-1/2013-14(BSR)/7-93/2007(BSR)).

A model for inter-laminar shear stress in laminated composites

W Zaki, V Nguyen and R Umer

Khalifa University, 127788 Abu Dhabi, UAE

wael.zaki@kustar.ac.ae

Abstract. The paper presents an analytical model for the estimation of shear stress at the interface of adhesively bonded layers in laminated composites. For this purpose, a new shear stress function is proposed that accounts for the influence of shear lag in laminates assembled using adhesive layers of different types and thicknesses. In addition to the estimation of interfacial stress, the function can be used to determine the difference in normal strains and axial displacements in adjoining layers.

1. Introduction

Composite laminates are used in a multitude of engineering applications because of the ability to achieve superior and directionally prevalent mechanical performance per unit weight compared to bulk materials [1]. In this regard, a key factor governing the overall mechanical performance of laminates is the efficiency of load transfer at the interface of adjoining layers, which is assumed to be perfect by conventional laminate theories. In reality, the transfer efficiency is affected by the development of interlaminar shear stress in the interface between these layers [2]. This shear stress is accounted for in this paper by means of a newly developed stress function. The parameters of the function are determined by fitting against numerical simulation results using finite element analysis. Experimental validation is then carried out for the case of a 3-layer adhesively bonded aluminium laminate.

2. Materials and methods

The model for interlaminar shear proposed in this work is derived based on the shear lag model (SLM) [3,4], which describes stress distribution and mechanical properties of the interface [5,6]. The model is developed for a laminate consisting of n metallic sheets, as shown in figure 1, each of width a, thickness t, and length $2L$. The metal sheets are bonded by means of adhesive layers of matching width and length, of thickness s.

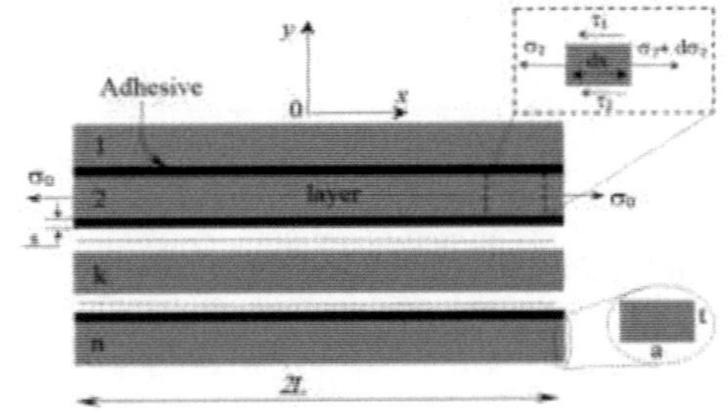

Figure 1. Adhesively bonded laminate.

IOP Conf. Series: Journal of Physics: Conf. Series **814** (2017) 012006 doi:10.1088/1742-6596/814/1/012006

The variation of the average normal stress $\sigma_{a.m}$ in the longitudinal direction x of layer m of the adhesive, where $m = 1$ to n, is related to the shear flow by means of the equation

$$\frac{d\sigma_{a.m}}{dx} = \frac{\tau_k - \tau_{k-1}}{t} = \frac{\tau_{m-1} - \tau_m}{t}$$

where $k = m - 1$ is the interface number.

The shear stress is proposed to be related to the difference in average normal strain ϵ_a in layers $m - 1$ and m as follows:

$$\tau_k = f(E_a) \cdot (\epsilon_{a.m-1} - \epsilon_{a.m})$$

where $f(E_a)$ is a function of the Young modulus E_a of the adhesive, which is assumed here to have the following expression:

$$f(E_a) = A \cdot e^{B \cdot E_a} + C \cdot e^{D.E_a}$$

in which A, B, C, D are scalar parameters that depend on the properties of the laminate. This expression of $f(E_a)$ is based on observations made in earlier work [2].

Once the parameters of the stress function $f(E_a)$ are determined, the distribution of shear stress in interface k can be found from the difference in normal strain $(\epsilon_{a.m-1} - \epsilon_{a.m})$. Moreover, if one side of the laminate is fixed, the average axial displacement of the tip in each layer m is obtained by the integral

$$u_{a.m} = \int_0^L \epsilon_{a.m} dx$$

2.1. Characterization of the parameters of the stress function

The general procedure for determining the parameters of the function $f(E_a)$ consists in fitting the analytical expression of the function to results of numerical simulation of the laminates. These results give access to numerical estimates of $f(E_a)$. As an example, finite element analysis of a 3-layer aluminium laminate with the parameters listed in Table 1 was carried out for adhesives with Young's moduli ranging from 0.9 to 5 GPa.

Table 1. Parameters used for finite element analysis of 3-layer aluminum laminates.

Parameter	Value
L	25 mm
t	1 mm
s	0.05 mm

In each case, the average tip displacement of the laminate was calculated and fitted to its expression in terms of $f(E_a)$, resulting in the values of A, B, C, D listed in Table 2.

Table 2. Parameters of the stress function for the simulated 3-layer aluminum laminate.

Parameter	Value
A	21.878 GPa
B	0.004969 GPa^{-1}
C	-18.43 GPa
D	-1.779 GPa^{-1}

A plot of the fitted function versus numerical simulation results is shown in figure 2.

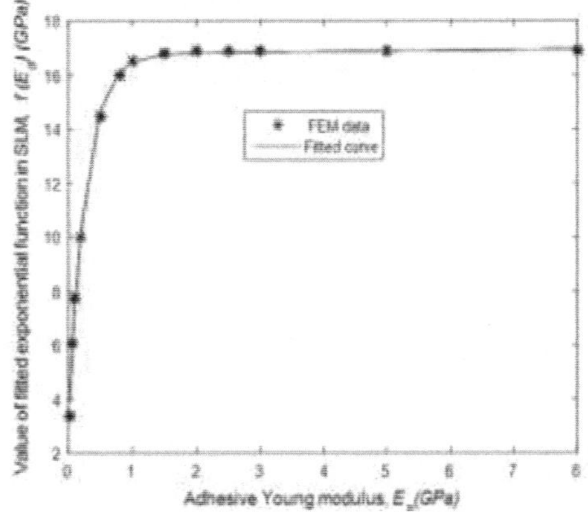

Figure 2. Fitting of the shear stress function to simulation results.

In this case, very good accuracy is achieved with a maximum relative error not exceeding 2% in magnitude (figure 3).

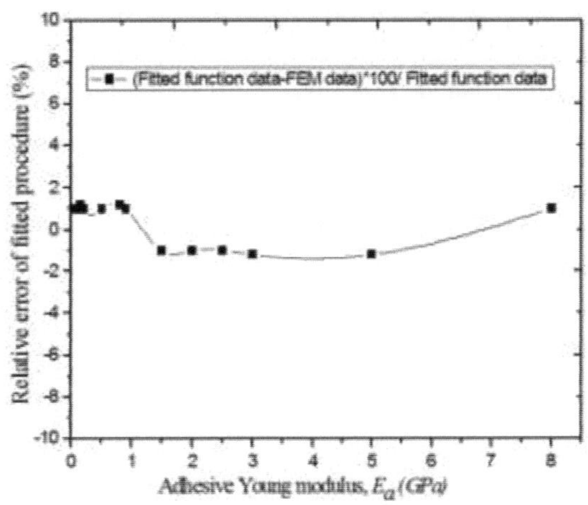

Figure 3. Relative error on the fitted stress function.

2.2. *Experimental validation*

Experimental validation was carried out for the case of 3-layer AA 1100 H-14 aluminum laminates, bonded using the Cytec PRISM EP2400 thermosetting epoxy resin. The aluminum layers consisted of plates of thickness t = 0.9 mm. The surfaces of the plates were abraded to uniform roughness resulting in a uniform mat appearance, then cleaned using alcohol and water in preparation for the application of the adhesive layer. The adhesive was then deposited uniformly on the surfaces to be joined, and the layers were clamped to avoid side-ways sliding. The resin was cured at 178 C for two hours then cooled and maintained at room temperature for 24 hours. The resulting Young modulus for the cured resin was $E_a = 3.4$ GPa according to manufacturer's specifications. The laminate was finally loaded in an Instron 5969 UTM equipped with a 50 kN load cell and the tip displacement measured using extensometers of gauge lengths 25, 50 and 75 mm.

As shown in Table 3, the experimental measurements are predicted to good accuracy by the analytical model, using the fitted stress function reported earlier.

Table 3. Tip displacement error for different gauge lengths.

	L_e=25 mm	L_e=50 mm	L_e=75 mm
Relative error %	10.9	11.9	14

3. Conclusion

A model for shear stress at the interface of adhesively bonded laminates was proposed based on SLM. The model introduces a novel analytical stress function to relate the shear stress that develops in an interface to the difference in axial strain, and therefore the elongation, in adjoining layers. The function is assumed to present exponential dependence on the Young modulus of the adhesive and features 4 scalar parameters that can be determined by fitting to numerical simulation results. Experimental validation of the proposed model was successfully carried out for the case of a 3-layer aluminum laminate bonded using a thermosetting resin. The work contributes to improving our understanding of the properties and behavior of adhesive interfaces and presents strong potential for application by simplifying the design and analysis of adhesively bonded composite laminates.

4. References

[1] Marimuthu R, Sundaresan M.K and Rao G.V 2003 *J. Inst. Eng.* 32–38.
[2] Viet N.V, Wang Q and Kuo W.S 2012 *Compos: Part B* **43** 332.
[3] Jiang G and Peters K 2008 *Int. J. Solids. Struct.* **45** 4049.
[4] Nairn J.A 2005 *Adv. Compos Lett* 06.
[5] Gao X.L and Li K 2005 *Int. J. Solids. Struct* **42** 1649–67.
[6] Her S.C, Liu S.J 2014 *Eng. Comput.* 353–364.

Acknowledgements

Dr. Wael Zaki would like to acknowledge the financial support of Khalifa University through KUIRF research grant no. 210114.

IOP Conf. Series: Journal of Physics: Conf. Series **814** (2017) 012007 doi:10.1088/1742-6596/814/1/012007

Comparison between a typical and a simplified model for blast load-induced structural response

A Abd-Elhamed[1,2] and S Mahmoud[3]

[1] Faculty of Engineering at Mataria, Helwan University, Cairo, Egypt
[2] Faculty of Engineering & Technology, Egyption Chinese University, Egypt
[3] Department of Construction Engineering, College of Engineering, University of Dammam, Saudi Arabia

E-mail: aymanm79@hotmail.com

Abstract. As explosive blasts continue to cause severe damage as well as victims in both civil and military environments. There is a bad need for understanding the behavior of structural elements to such extremely short duration dynamic loads where it is of great concern nowadays. Due to the complexity of the typical blast pressure profile model and in order to reduce the modelling and computational efforts, the simplified triangle model for blast loads profile is used to analyze structural response. This simplified model considers only the positive phase and ignores the suction phase which characterizes the typical one in simulating blast loads. The closed from solution for the equation of motion under blast load as a forcing term modelled either typical or simplified models has been derived. The considered herein two approaches have been compared using the obtained results from simulation response analysis of a building structure under an applied blast load. The computed error in simulating response using the simplified model with respect to the typical one has been computed. In general, both simplified and typical models can perform the dynamic blast-load induced response of building structures. However, the simplified one shows a remarkably different response behavior as compared to the typical one despite its simplicity and the use of only positive phase for simulating the explosive loads. The prediction of the dynamic system responses using the simplified model is not satisfactory due to the obtained larger errors as compared to the system responses obtained using the typical one.

1. Introduction

Accidental explosions and deliberate bombings by terrorists cause catastrophic structural damage which results in downtime such structures and create a negative impact on society. This phenomenon may lead to severe damage or even damage to numerous surrounding structures. Moreover, loss of life can result from the collapse of structures [1-4]. Several research works have investigated the different models of air blast and the effect of air blast loads on reinforced concrete structures. The evolution of blast pressure with time can be simulated either by exponential distribution, which depends on explosive charge size, type, and distance to a target, or by triangular distribution through neglecting the suction stage which leads to the simplified model [5-7].

The methodologies available nowadays for prediction of blast effects on building structures can be divided into experimental, analytical and numerical procedures. Although the experimental tests are unsafe, expensive and dangerous, they can provide a wealth of useful data.

IOP Conf. Series: Journal of Physics: Conf. Series **814** (2017) 012007 doi:10.1088/1742-6596/814/1/012007

The theoretical approach usually utilizes idealized models and building structures modelled as single-degree-of freedom systems (SDOF). Although this procedure is based on several assumptions and approximations, it is considered as the most applicable one and rather applicable in routine design where it provides researchers and structural designers with useful data and results [8]. This procedure substitutes the structural element by an equivalent stiffness, SDOF structural system and an elastic-plastic response spectra to capture the peak responses of modelled system. With the development of computer technology, a lot of researchers paid attention for conducting research works concerning the blast induced structural response numerically. The equations of motion are widely used to describe complex models in various field of engineering, particularly in structural dynamic. The wide applications of these equations are the main reason behind attracting mathematicians for performing blast analysis in the last decades. In order to understand the physical mechanism of phenomena in nature described by second order non homogeneous ordinary differential equation, exact solutions have to be explored. The study of these equations becomes one of the most important topics in mathematical physics.
This research aims to drive the exact solution of the dynamic response of buildings under blast loads for two types of pressure configurations, more accurate exponential distribution of blast pressure and simplified triangular blast pressure.

2. Modelling and idealization

The structure under study is one which can be represented by idealized SDOF mathematical model for predicting dynamic response of concrete structures subjected to blast loading as shown in figure 1. A rigid deck is connected to a base through massless columns of stiffness *k*, representing the resistance of structure against deformation, and damping coefficient *c*. The values of structural stiffness and damping coefficients can be calculated from the formulas [9].

$$k = \frac{4\pi^2 m}{T^2}; \quad c = 2\xi\sqrt{km} \tag{1}$$

where T, and ζ denote the natural structural vibration period and structural damping, respectively.

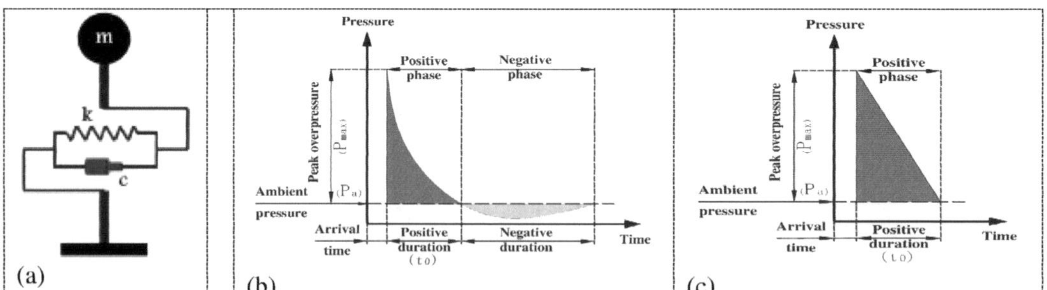

Figure 1. (a) Idealized SDOF building model (b) typical blast load profile (c) simplified blast load profile.

The equation of motion of the idealized building model presented in figure 1(a) and subjected to a blast pressure $P(t)$ can take the form:

$$m\ddot{u} + c\dot{u} + ku = A_{eq}\, p(t) \tag{2}$$

where u, $\dot{u}$, *and* $\ddot{u}$ represent the displacement, velocity, and acceleration of the superstructure respectively, while, A_{eq} is the exposed surface area of the building model. The blast pressure distribution $p(t)$ as function of time t, the positive phase duration t_0, and the peak overpressure P_{max} can take either the typical form (figure 1(b)):

$$p(t) = p_{max}(1-\frac{t}{t_0})\exp\left(\frac{-bt}{t_0}\right)A_{eq}. \tag{3}$$

Or the simplified form (figure 1(c)):

$$p(t) = p_{max}(1-\frac{t}{t_0}) \tag{4}$$

where b is a shape parameter depending on the dimensionless characteristic scaled distance Z [5] as:

$$Z = \frac{R}{W^{1/3}} \tag{5}$$

where W is an explosive charge, expressed in Kilograms of TNT, and R is the standoff distance, or distance between a point of interest and the blast epicenter.

3. Closed form solution

Case 1: equation of motion with forcing term modeled as typical model

Referring to equations (2) and (3), the equation of motion of the building model under typical blast pressure is

$$m\ddot{u} + c\dot{u} + ku = A_{eq}p_{max}(1-\frac{t}{t_0})\exp\left(\frac{-bt}{t_0}\right) \tag{6}$$

With initial conditions:

$$u(0) = 0, \ \dot{u}(0) = 0 \tag{7}$$

Equation (6) together with initial conditions in equation (7) can be solved analytically in terms of the particular and homogenous solutions to determine the time-dependent displacement, velocity, and acceleration of the system following any of the available textbooks [10] as:

$$u_p = (A + Bt)\exp\left(\frac{-bt}{t_0}\right) \tag{8}$$

where u_p is the particular solution. A and B are constants to be determined applying the method of determined coefficients as:

$$B = \frac{-p_{max}A_{eq}}{t_0(m\frac{b^2}{t_0^2} - c\frac{b}{t_0} + k)} \tag{9}$$

$$A = -Bt_0 + \frac{B^2}{p_{max}A_{eq}}(c - 2m\frac{b}{t_0})t_0 \tag{10}$$

The homogenous solution u_h can be written as:

$$u_h = \exp(\alpha t)[c_1\cos\beta t + c_2\sin\beta t] \tag{11}$$

where c_1, c_2, α and β are another constants to be determined employing the characteristic equation and the initial conditions.

Applying the characteristic equation α and β have been found in terms of mass, stiffness and damping coefficient of the building model as:

$$\alpha = \frac{-c}{2m}, \quad \beta = \frac{(c^2 - 4mk)^{1/2}}{2m} \tag{12}$$

Knowing that the general solution u is the summation of both homogeneous and particular solution:

$$u = u_h + u_p \tag{13}$$

$$u = \exp(\alpha t)\left[c_1 \cos\beta t + c_2 \sin\beta t\right] + (A + Bt)\exp\left(\frac{-bt}{t_0}\right) \tag{14}$$

Derivative of u with respect to time t can be derived as:

$$\begin{aligned} \dot{u} &= \alpha \exp(\alpha t)\left[c_1 \cos\beta t + c_2 \sin\beta t\right] + \exp(\alpha t)\left[-c_1\beta \sin\beta t + c_2\beta \cos\beta t\right] \\ &+ B \exp\left(\frac{-bt}{t_0}\right) - \frac{b}{t_0}\exp\left(\frac{-bt}{t_0}\right)(A + Bt) \end{aligned} \tag{15}$$

Similarly, derivative of $\dot{u}$ with respect to time t can be derived as:

$$\begin{aligned} \ddot{u} &= \alpha^2 \exp(\alpha t)\left[c_1 \cos\beta t + c_2 \sin\beta t\right] + 2\alpha \exp(\alpha t)\left[-c_1\beta \sin\beta t + c_2\beta \cos\beta t\right] \\ &- \exp(\alpha t)\left[c_1\beta^2 \cos\beta t + c_2\beta^2 \sin\beta t\right] - 2B\frac{b}{t_0}\exp\left(\frac{-bt}{t_0}\right) + \frac{b^2}{t_0^2}\exp\left(\frac{-bt}{t_0}\right)(A + Bt) \end{aligned} \tag{16}$$

where A, B, α and β determined as before.
In order to find the constants c_1 and c_2 one needs to apply initial conditions to equations (14) and (15).

$$c_1 = -A \tag{17}$$

$$c_2 = \frac{A(\alpha + \frac{b}{t_0}) - B}{\beta} \tag{18}$$

Case 2: equation of motion with forcing term modeled as simplified model
Referring to equations (2) and (4), the equation of motion of the building model under simplified blast pressure is

$$m\ddot{u} + c\dot{u} + ku = p(t) \tag{19}$$

where

$$P(t) = \begin{cases} P_{\max}(1 - \frac{t}{t_0})A_{eq} & t \le t_0 \\ 0 & t \ge t_0 \end{cases} \tag{20}$$

With the same initial conditions mentioned in equation (7).

For the case $t \le t_0$:
The particular solution can take linear form as:

$$u_p = A + Bt \tag{21}$$

Obtaining the derivatives of equation (16) and substituting into equations (14) and comparing coefficients:

$$B = -\frac{P_{\max}}{kt_0} A_{eq} \tag{22}$$

$$A = -\frac{P_{\max} A_{eq} (kt_0 + c)}{k^2 t_0} \tag{23}$$

Substituting the values of A and B into equations (16)

$$u_p = -\frac{P_{\max} A_{eq}}{kt_0}(t_0 + \frac{c}{k} - t) \tag{24}$$

Similarly to the typical model case the homogeneous solution can be found as|:

$$u_h = \exp(\alpha t)[c_1 \cos \beta t + c_2 \sin \beta t] \tag{25}$$

The values of α and β can be calculated as in equation (12).

In order to find the constants c_1 and c_2 one needs to apply initial conditions to the general solution equation.

Knowing that the general solution u is the summation of both homogeneous and particular solution:

$$u = u_h + u_p \tag{26}$$

$$u_h = \exp(\alpha t)\left[c_1 \cos \beta t + c_2 \sin \beta t\right] + \frac{P_{\max} A_{eq}}{kt_0}(t_0 + \frac{c}{k} - t) \tag{27}$$

Derivative of u with respect to time t can be derived as:

$$\dot{u} = \exp(\alpha t)\left[-c_1\beta \sin \beta t + c_2\beta \cos \beta t\right] + \alpha \exp(\alpha t)\left[c_1 \cos \beta t + c_2 \sin \beta t\right] - \frac{P_{\max} A_{eq}}{kt_0} \tag{28}$$

Similarly, derivative of $\dot{u}$ with respect to time t can be derived as:

$$\begin{aligned} \ddot{u} &= \alpha \exp(\alpha t)\left[-c_1\beta \sin \beta t + c_2\beta \cos \beta t\right] + \exp(\alpha t)\left[-c_1\beta^2 \cos \beta t - c_2\beta^2 \sin \beta t\right] \\ &+ \alpha^2 \exp(\alpha t)\left[c_1 \cos \beta t + c_2 \sin \beta t\right] + \alpha \exp(\alpha t)\left[-c_1\beta \sin \beta t + c_2\beta \cos \beta t\right] \end{aligned} \tag{29}$$

Now we apply initial conditions to the general solution equation (27) and equation (28).

$$c_1 = -\frac{P_{\max} A_{eq}}{kt_0}(t_0 + \frac{c}{k}) \tag{30}$$

$$c_2 = \frac{P_{max} A_{eq}}{\beta kt_0}[1 + \alpha(t_0 + \frac{c}{k})] \tag{31}$$

For the case $t \geq t_0$***:***

The forcing term $P(t)=0$, the general solution $u_{t \geq t_0}$ can be expressed as:

$$u_{t \geq t_0} = \exp(\alpha t)\left[G_1 \cos \beta t + G_2 \sin \beta t\right] \tag{32}$$

Derivative of $u_{t \geq t_0}$ with respect to time t

$$\dot{u}_{t \geq t_0} = \exp(\alpha t)\left[-G_1\beta \sin \beta t + G_2\beta \cos \beta t\right] + \exp(\alpha t)\left[G_1 \cos \beta t + G_2 \sin \beta t\right] \tag{33}$$

Derivative of $\dot{u}_{t \geq t_0}$ with respect to time t

$$\begin{aligned}\ddot{u}_{t \geq t_0} &= \exp(\alpha t)\left[-G_1\beta^2 c\,os\beta t - G_2\beta^2 \sin\beta t\right] + \alpha \exp(\alpha t)\left[-G_1\beta \sin\beta t + G_2\beta \cos\beta t\right] \\ &+ \alpha^2 \exp(\alpha t)\left[G_1 \cos\beta t + G_2 \sin\beta t\right] + \alpha \exp(\alpha t)\left[-G_1\beta \sin\beta t + G_2\beta \cos\beta t\right]\end{aligned} \quad (34)$$

Appling the initial conditions (as illustrated in figure (2)):

$$u_{t \geq t_0} = u_{t = t_0} \text{ and } \dot{u}_{t \geq t_0} = \dot{u}_{t = t_0} \quad (35)$$

Leads to:

$$G_1 = \frac{R - G_2 \sin(\beta t_0)}{\cos(\beta t_0)} \quad (36)$$

$$G_2 = \frac{E - R\left(\alpha - \beta \tan(\beta t_0)\right)}{\beta \cos(\beta t_0)\left(1 + \tan^2(\beta t_0)\right)} \quad (37)$$

where $$R = u(t_0) / \exp(\alpha t_0) \quad (38)$$

and $$E = \dot{u}(t_0) / \exp(\alpha t_0) \quad (39)$$

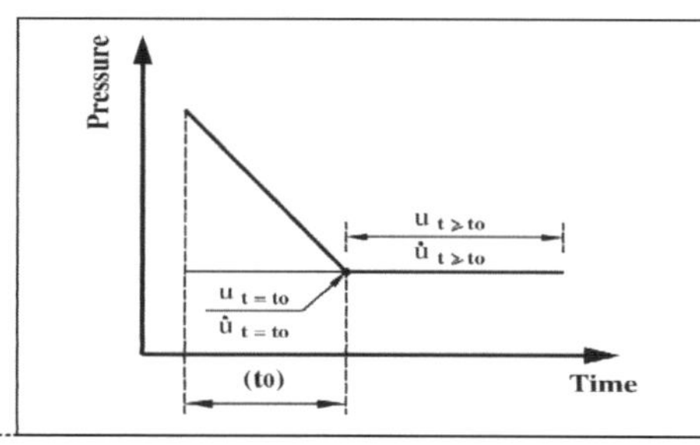

Figure 2. Initial conditions for the case t ≥ t_0.

Figure 2 presents the simplified triangle distribution for the blast load as well as the initial condition required to get the exact solution of equation (19).

4. Results and discussion

To demonstrate the validity of the simplified blast load model, the closed form solution of the equation of motion for the building model shown in figure 1(a) using the two models of blast loads presented in figure 1(b) and (c) are provided using an explosive load of peak pressure P_{max} of 312 kpa and positive phase duration t_0 = 0.02142 second. The considered system parameters in terms of mass m, damping ratio ζ and natural period T_n, are set to be 25 $\times 10^3$ kg, 0.05 and 1.2 s respectively.
Here, for brevity, a comparison between typical and simplified blast pressure profile models is performed.

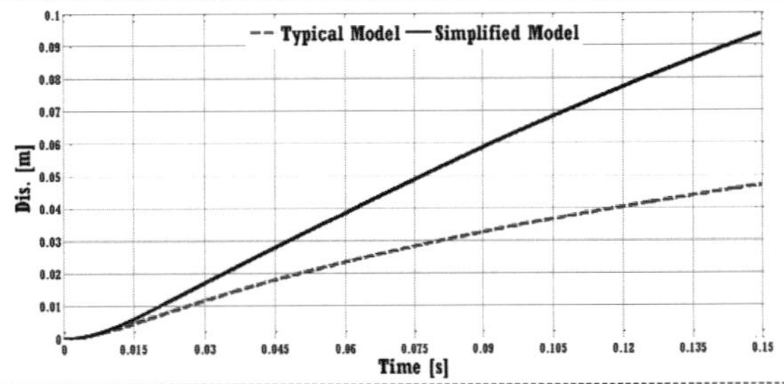

Figure 3. Storey displacement time histories with two models of blast load.

It can be seen from figure 3 that the displacement of the building model reaches its peak at the end of explosion time. Moreover, it has been noticed that, the error in predicted displacement response increases with increasing the blast duration.

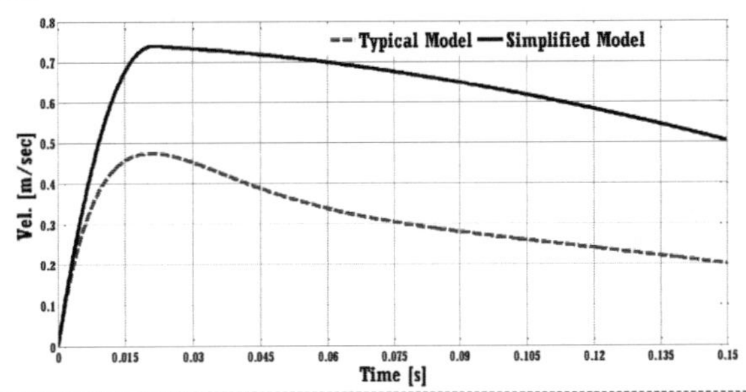

Figure 4. Storey velocity time histories with two models of blast load.

Figure 4 presents the captured velocity time-histories of the considered building model using the typical and simplified blast models. It is worth noting that the velocity of the structure increases rapidly in the initial time *to* reach its peak value almost at the end of the positive phase duration $t = t_0$, and the corresponding error reaches maximum at the end of positive phase as well.

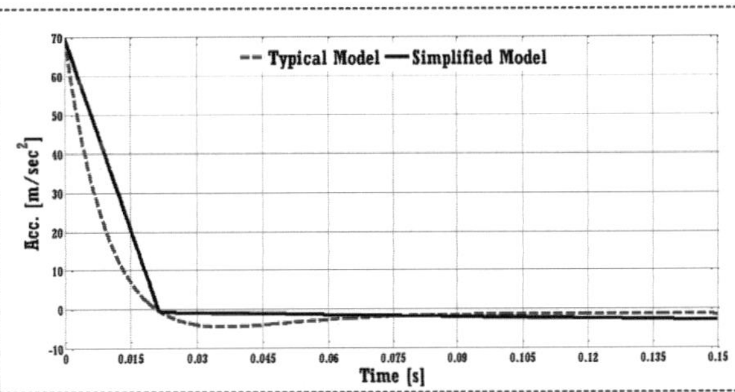

Figure 5. Storey acceleration time histories with two models of blast load.

Focusing on the acceleration curves shown figure 5, it can be seen that the accelerations start with certain maximum values immediately at the occurrence of the blast detonation followed by a sudden decrease to zero for the remaining time duration. Therefore, the acceleration response appears to be only significant at the start of the explosion causing sudden increase in the storey accelerations. Moreover, it has been noticed that, the obtained peak acceleration response with simplified model is nearly identical to the one obtained with typical model.

The captured peak displacement and velocity for the considered herein building model subjected to the aforementioned blast load in the form of typical model are 0.0469 m, and 0.4748 m/s respectively. Considering the blast load in the form of the simplified model one increases these captured values to be 0.0935 m, and 0.7396 m/s respectively. Consequently, the simplified model overestimates the peak responses by 100% and 60% respectively. But, the peak acceleration remains constant and equal 69.7986 m/sec^2 in both of the two models.

As it can be seen from the presented figures and the obtained results, the proposed simplified model excessively increases the obtained response for the building model during the whole period of explosion and does not yield good predictions except for the peak acceleration response.

5. Conclusions

This paper presents a comprehensive investigation on the induced structural response due to applied blast loads. Both typical and simplified models have been used to represent the explosive load acting on the building model. It has been demonstrated that the simplified model for blast load yields unreasonable predictions of structural responses as compared to the typical model where a significant difference between the induced structural responses of the buildings has been found.

References

[1] Osteraas JD 2006 Murrah building bombing revisited: A qualitative assessment of blast damage and collapse patterns *J. Perform. Constr. Facil.* **20** pp 330–335.

[2] Elliot CL, Mays GC and Smith PD 1992 Proceedings of Institution of Civil Engineers, Structures & Buildings, **94** pp 287–297

[3] Corley W G 2004 J. Perform. Constr. Fac., **18** pp 68-78

[4] Remennikov A M 2003 J. of Battlefield Tech. Aragon Press Pty Ltd., **6** pp 5-10

[5] Baker W.D., Cox, P.A., Westine, P.S., Kulesz, J.J. and Strehlow, R.A., 1983 Explosion Hazards and Evaluation, Elsevier, Amsterdam, the Netherlands

[6] Kinney GF, Graham KJ 1985 Explosive Shocks in Air, Springer-Verlag, New York

[7] Bangash MYH 1993 Impact and Explosion, Analysis and Design, Blackwell Scientific

[8] Mays GC and Smith PD 1995 Blast Effects on Buildings, Thomas Telford Publications, Heron Quay, London

[9] Harris CM and Piersol AG 2002 in: *Harris' Shock and Vibration Handbook*, edited by Harris CM, and Piersol AG, McGraw-Hill publishing, New York

[10] *Zill* DG 2001 *Differential Equations with Modeling Applications*, 7th ed., Brooks/Cole Thompson Learning, Pacific Grove, CA

IOP Conf. Series: Journal of Physics: Conf. Series 8142017) 012008 doi:10.1088/1742-6596/814/1/0120C

The research of 4th grade mathematical curriculum electronic picture book construction and development in integrating indigenous culture

Yen Ting Chen[1], Juei Hsin Wang[2*]

[1]Department of Mathematics Education, National Taichung University of Education, Taiwan
[2*]Graduate Institute of Educational Administration and Policy Development, National Chiayi University, Taiwan

E-mail: ytchen@mail.ntcu.edu.tw; gloriawang2004@mail.ncyu.edu.tw

Abstract. This research aimed at integrating Seediq culture and mathematical course design for fourth-grade elementary school, and then transforming this mathematical course into an electronic picture book. During the process of electronic book development, the researchers collected videos of six participants engaged in discussion, reflection minutes after the meeting written by the attendants, the researchers' observation and review journals, and conversations with the participants. Then, researchers utilized Content Analysis to explore, try, review and retry steps of electronic book making process. The main findings: There are four periods of electronic book making process, research occurrence period, curriculum design period, electronic book transformation period, and result evaluation period. The picture book included the White Stone Legend born from Seediq seniors, historical battle for hunting field between tribes, and concepts of approximation, angle, triangle, and quadrangle features. At last, with the research result, this article presents the corroboration of related works, and then proposes suggestions of electronic book teaching and follow-up studies.

1. Introduction

In the past, a lot of research [1, 2, 3] has indicated that the teaching activity design in normal mathematical classrooms rarely pay attention to the learning of underprivileged students and the students from minority ethnic groups. Hence, some countries (America, Canada, and Australia) actively participate in making materials [1, 2, 3, 4] suitable to students' basic knowledge and culture. Besides, Ministry of Education emphasized, in general program of 12-year compulsory education, on curriculum demands considering multicultural and ethnic diversity. Earlier studies [2] have also mentioned that native indigenous students have no interest in learning math and show poor academic performances. targeting the dominant culture might result in such situations [1]. Accordingly, designing mathematical materials, which is based on the prior knowledge of the students from minority ethnic groups and their cultural, is a task worthy putting effort on.

However, looking around native 16 indigenous groups, Seediq was recognized as the 13th indigenous group officially in 2008. It was relatively a new group identified by the Indigenous Committee. There are 48 essays focusing on Seediq culture in the National Library theses and dissertations system. There are total 43 theses focusing on Seediq culture in journal theses system, and only two of them

considered teaching to Seediq students [5]. Obviously, it is highly important to consider culture while developing math curriculum for Seediq students.
Furthermore, most students like picture books with plots [6]. Reading illustrations and words in a book is as interesting as watching dramas [7]. Employing picture books in education improves learning of new knowledge and language, enhances reading ability, and cultivates creativity [8]. Yet, as technology developed, the Ministry of Education mentioned that we should understand the influences of science and techniques to humans, and learn how to utilize and manage science as well as skills in Grade 1-9 (Curriculum Guidelines : Science and Technology learning area). Electronic books are the emerging multi-media products in this context. Teaching through electronic books not only stimulates students ' visuals, which help them in learning math actively [1], but also makes them learn faster [9]. Combing picture and electronic books and equipping them with sound, voice and, animation transforms the static learning system into a dynamic one. It also works as an appropriate teaching material, which is responsive and has immediate implications as per the learning schedule and students' demands [10].
In summary, this study focusing on Seediq culture attempted to develop mathematical learning materials (this research utilized the myth: White Stone Legend and the history of ethnic conflict) for Seediq children in the 4th grade and added concepts of approximation (4-n-06), angle (4-s-01, 4-s-04, 4-s-05), triangle, and quadrangle (4-s-07). Then, a mathematical learning curriculum focusing on indigenous culture was developed, which is then transformed into an electronic picture book. Through the process of electronic picture book development, we aimed to determine the significant of this process in the math learning among 4th grader Seediq students.

2. Theoretical foundations

At present, the process of learning and teaching emphasizes the importance of "reflection". Reflection is that a learner reflects in action, which is the key to learn [11]. In order to abound mathematical learning materials for Seediq students, this research considered act of researchers as well as six students aspiring to become teachers. During the process, the researchers reflected on existing problems and proposed certain problem solving strategies, which became the research aim in the next period and was then carried out. Furthermore, the researchers found new problems, reflected, and then purposed new problem solving strategies and action project, while carrying out the project. The strategies mentioned above were consistent with the group cooperation, conversation, judgments, reflection and problem solving direction in action research [12].

3. Research design

3.1. Research situation

With a goal of developing mathematical curriculum electronic picture book based on Seediq culture for 4th grade elementary students, the researchers recruited six students undergoing teacher- training from their working school, who were interested in learning of indigenous students with hope to improve their mathematical teaching knowledge to carry out the six-month research and development process. The researchers utilized their past mathematical curriculum experience based on indigenous culture and suggested six steps for the electronic book development: culture analysis, curriculum analysis, story context, mathematical problem production, peer judgments and revision, picture book painting, and electronic producing. The given figure 1 and Table 1 explain each step.

3.2. Research participants

Among the six students in teacher training, one (Xuan) is the second grade college student of Cultural and Creative Industry Design and Management Department, and the other five (Han, Cheng, Hao, Qiao, Jay) are college students of Mathematical Education Department (four juniors and one freshman). During the past year, they followed researchers in integrating Tsou culture with mathematical curriculum design hence they were having some related experiences.

3.3. Research tools: electronic book producing tool

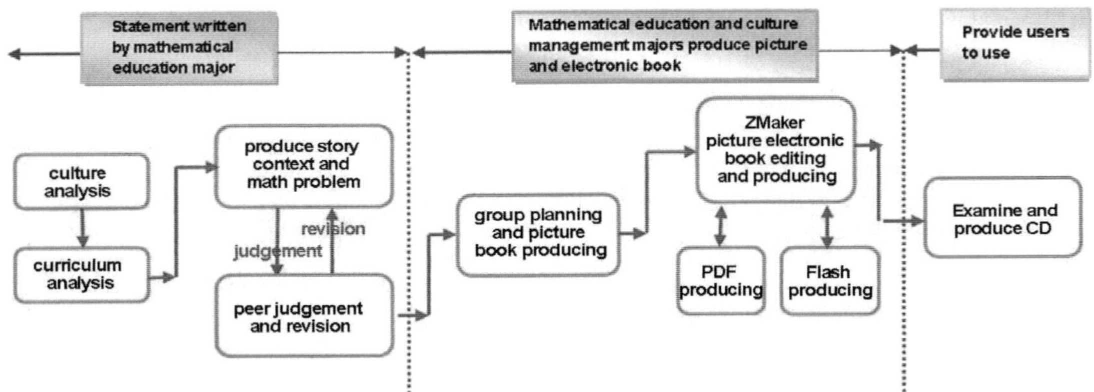

Figure 1. Operating procedure of picture electronic book producing

Table 1. Each steps and detail in this research

step	time	detail explanation
1. culture analysis	103.12	1-1. collect and read literature related to Seediq 1-2. visit Seediq tribe in person 1-3. visit school teachers in Seediq tribe
2. curriculum analysis	104.01	2-1. analyze context in mathematical textbook of 4th grade student 2-2. understand 4th grade students' mathematical prior knowledge
3. story context and mathematical problem construction	104.02	3-1. add Seediq culture into story plot 3-2. add mathematical conception and problem in story plot
4. peer judgement and revision	104.03	4-1. encourage teachers in tribe to proofread story and mathematical question 4-2. revise work according to suggestions provided by tribe teachers
5. paint picture book	104.04	5-1. design pictures and discuss 5-2. color pictures and typesetting
6. produce picture electronic book	104.05	6.1 seek multi-media (voice、animation) to upgrade the value 6-2. electronic page editing and producing

The software chosen to produce electronic picture book was Zmaker, electronic book producing master software, which is easy to learn for a new user. Therefore, the researchers chose this software to produce electronic picture book.

3.4. Data collection and analysis

The collected data included weekly conversation videos (25 minutes in total), reflection minute written by the six students in teacher training after every meeting (150 in total), and the researchers' observation and review journals after meeting (25 in total). For this data analysis, the researchers utilized "Analysis Induction" because all data collected in this research were of qualitative nature. In the process of electronic book development, the researchers had a close look at the data in every category point by point in the line of "time", cross-compared the qualitative content of the participants' conversation and words in the way of induction and analysis, found problems emerging during a period with the help of the verification of "reflection journals" written by the researchers, and made "action project" to reflect and solve problems as well as practice the situation.

3.5. Research authenticity examination

Authenticity of this research was validated by 3-side cross examination of the "different data" (for example: conversation video, reflection minute and reflection journal after a meeting) and "different participants". One researcher (R1) with others having PhD degree was mainly in charge of the 3-side cross examination of "different participants"; they picked one ten corresponding to each other.th of the data randomly to categorize them

4. Research result and discussion

At the beginning of the electronic picture book development, the researchers foresaw difficulties during the research due to the lack of a previous design presenting integration of indigenous culture with mathematical curriculum. Hence, with the spirit of research, they tried to report problems emerging during the process of electronic book development, reflected on problems faced, and made a new action project to solve the problems.

4.1. Problem analysis and new action project

4.1.1. Problem analysis I. This study considered theSeediq culture in all aspects including cuisine, clothing, architectures, mythology, and legends. Therefore, the researchers decided to embark both the aspects of the culture: the "static: ancient books and historical records" and the "dynamic: visiting school teachers in the village". They found that "much information about the Seediq culture is not available on the Internet or in the libraries (J 1-03)".

*4.1.2. New action project I. After thinking carefully, the researchers decided to turn to a professor. The professor was in contact of several students, and was under the charge of principals and directors in Nantou County. Then it was possible for him to find those who are Seediq. As expected, they ended up finding two Seediq teachers. "**They saw many architectures and utensils through the lead by the two teachers** (M2- Hao-05)". The researchers found "**school teachers who can offer consultation in the tribe through connection, but they still hoped to seek more cultural information different from the oral history or entity** (J 2-15)".*

4.2. Problem analysis and new action project in picture book design period

*4.2.1. Problem analysis II. After obtaining enough cultural data, the researchers started cultural analysis. However, the analysis of the Seediq cultural objects. (clothes, houses, and rice mashers, see figure 2) found that the mathematical concepts hidden in them were most geometrical. "**I found that most mathematical concepts hidden in these houses and clothes are related to a plane figure** (M 3-Jay-18)." "**I found that the shapes of wooden houses and stone houses mostly are a kind of rectangle and parallelogram as well** (M 3-Hao-20)."*

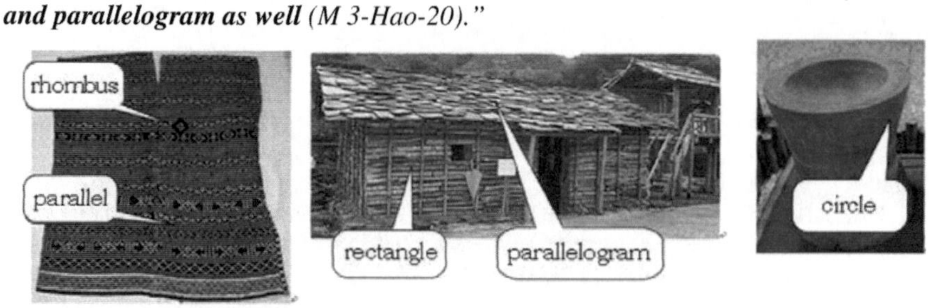

Figure 2. Seediq objects

*4.2.2. New action project II. Regarding the problems in the process of culture analysis, the researchers asked: "**According to van Hiele's geometry cognition level, these concepts superficially seemed plane geometry only and should be elevated to geometry quality analysis level** (J3-13). Therefore, the students in the teacher-training meditated, "**These cultural objects mostly belong to geometry concepts. Apart from knowing geometrical figures, are there other ways to elevate the problem level and inspire the children to think?** (M 4-T-27)." With the help of this guide, one student got an idea that "Seediq ethnic group has White Stone Legend and some battles with other ethnic groups: "**We can combine them into a story then think mathematical problems related to the story** (M 4-Qiao-35)".*

*4.2.3. Problem analysis III. However, after meditation and construction of trial problem, a researcher said, "**The problem construction this time was too similar to handouts. Problem construction in this way was no different than the general handouts** (M 5-T-38)". "**Take Jay's problem for example: There are 124 Taroko adults. Problem: How many people are there after approximating to the nearest tenth? This was alike the problems in textbooks, and the only difference is the addition of Taroko people** (M 5-T-39)", a teacher said. Besides, Han also said, "**I have a difficulty while designing problems, which is hard to use two concepts in one situation** (M5-Han-42)." This difficulty also exists in Hao's reflection: How to combine two math concepts in one context? It is a little bit difficult. "**It seems hard to enter geometry and other themes if the situation is number and calculation** (R5-Hao-07)".*

4.2.4. New action project III. Having difficulties as mentioned above, the researcher guided their thinking: "Another path of teaching is to make students learn math concepts through "doing by hands". The current math teaching focuses on ability cultivation. "We should cultivate children's ability of speaking reasons and make them express their thoughts (M 6-T-15~17)." Structure of culture integrating mathematical context is shown in figure 3.

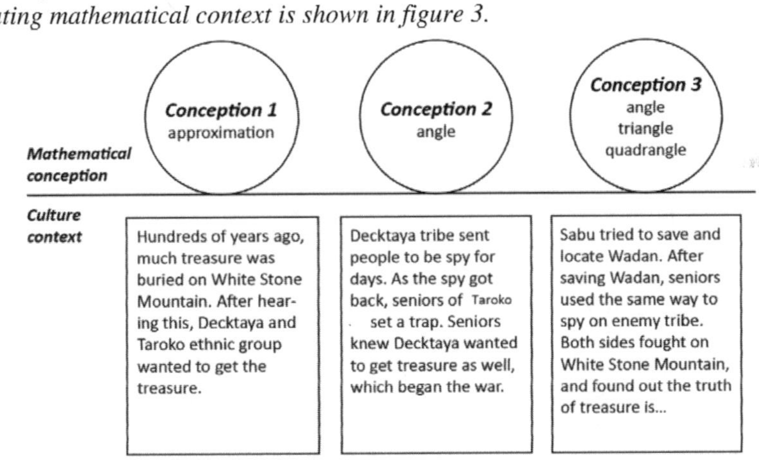

Figure 3. Structure of culture integrating mathematical context

Then, people discussed about the problems designed in this structure. Take Hao's problem for example: Sabu shoots birds in the sky. Problem: How many degrees is the elevation angle of this arrow shot? After discussion, it became a problem for verification (figure 4). The researchers judged the design of the problems, "after this discussion, we came up with many different thoughts and changed most common mathematical problems into problems belonging to the verification (R 8-Cheng-05). "After a brainstorm, the standardized mathematical problems are verified making them more and more flexible… (J 13-07)", a researcher noted. By peer judging and revising the problems one by one, the group finished revising all the mathematical problems in the cultural context.

4.3. "Problem Analysis" and the "New Action Project" in electronic book transform period

In the next step, the group was ready to produce an electronic book. First, the participants had a discussion about the drawing idea. "*As for the drawings in picture books are concerned, after a discussion, we decided it is all up to Xuan* (M 14-Han-05~06)." *It turns out that "...according to my impression, indigenous people have rectangular faces; I therefore, draw so* (M 14-Xuan-09)." The case proves that "*the characteristic of the roles would be subtly embedded with the drawer's impression of indigenous people* (J 14-06) (figure5)." But Jay had a different idea of it: "I think it is okay to make a cute version, which is more attractive to students (M14-Jie-18)."

Figure 4. Problem of verification

Figure 5. Student involved in this research

4.3.1. Problem analysis IV. In response to the question from the researcher, Hao said, "... in position of readers, how we can tell it as Jimmy's work, once seeing his drawings? It's because of his unique drawing style. Therefore, I support Xuan for keeping her unique style (M 15-Hao-09~11)." By digging into the spirit of picture books and self-reflection, the participants reach the decision of "respecting the original ideas", which make the figure of characters as most vivid feature of the picture book.

4.3.2. Problem analysis V. The following step is about feeding the drawings on paper into electronic book producing software, Zmaker. "... At the beginning, the electronic book presents the idea of our design, helping readers know the concept with which we design the picture book and the grade and the units, it suits (M 16-Hao-05).

4.3.3. New action project V. Conceding the lack of sound effect and animation in the electronic book as purposed by the peers, Han decided to be in charge of animation production. As for sound effect, Qiao suggested "It should be some different human voices to cooperate with one another (M 17-Qiao-12)", which got much recognition.

5. Conclusion and suggestions

In conclusion, this research was aimed at designing an electronic picture book integrating Seediq culture with mathematical curriculum for 4th grade elementary school. Through the cyclic process of action research "problem analysis → reflection → constructing and carrying out new action project", the research group overcame the difficulties associated with this rare culture, incapability of finding corresponding mathematical concepts in the culture, picture book production, an electronic picture book equipped with animation and sound effects, and finally completed the electronic picture book. Besides, with the help of teaching demonstration in person, the researchers ameliorated the tribal teachers' worries that "schedule would be delayed due to this" and "students would learn less due to this", then successfully persuaded the teachers to teach with electronic picture book in person. Finally, the tribe teachers realized that this electronic book, integrating culture and mathematical concepts, does improve learning motivation among students and gives them a sense of familiarity in mathematical problem solving.

From Seediq cultural context, this research combines White Stone Legend and the battle between the two ethnic groups to create a story for this research. Next, researchers sought for possible mathematical concepts embedded in this cultural context and developed mathematical problems within the context. Then following this situation, they painted picture book on paper and finally produced multifunctional electronic book with animation, voice, and interaction, which is suggestive for future design and mathematical electronic media for indigenous people.

Moreover, on the way of execution of this research, we faced several "problems" during the development of the electronic picture book, such as difficulty in cultural data access, incapability of combining high-level mathematical concepts, and no differentiation between mathematical problems that we created and those already in textbooks, etc. However, making good use of connection and interaction and reflection by the peers finally ended the predicament in the developmental process, and resulted in satisfaction of the participants. These problem analysis and new action projects can be crucial references for electronic picture book integrating indigenous culture and mathematical concepts in the future.

Although mathematical picture book integrated with indigenous culture produced in this research gets recognition of the tribe teachers, deeply experiencing and understanding indigenous culture is important and inevitable challenge for Han Chinese designers. To conquer this challenge, the researchers suggest gathering more cultural data and seeking consultation from seniors, organizations and staff related to Council of Indigenous Peoples, which improved the development process of electronic picture book. Of course, evaluation tools designed by teachers for electronic book related mathematical concepts is also an important factor to affect teachers' willingness of using such an outcome. The researchers also recommend performing pretest and posttest design evaluation and believe that teachers will be more willing to teach with electronic picture books.

6. References

[1] Ru-Fen, Yao. (2014). When Math Meets Indigenous Culture: A Case Study of Developing Mathematics Modules for Indigenous Students. Chinese Journal of Science Education, 22(2), 135-161.

[2] Chih-Hsien, Huang., and Fou-Lai, Lin (2008). The Analysis of the Learning of Taiwanese Tayal Junior High School Students and the Design of Teaching Activities Utilizing Activity Theory. Chinese Journal of Science Education, 16(2), 147-169.

[3] Baxter, J. A., Woodward, J., Olson, D. & Robyns, J. (2002). Blueprint for writing in middle school mathematics, Mathematics Teaching in the Middle School, 8(1), 52-56.

[4] Empson, S. B. (2003). Low-performing students and teaching fractions for understanding: An interactional analysis. Journal for Research in Mathematics Education, 34(4), 305–343.

[5] Chih-jung, Liu (2012). The Effect Of Popular Science Activities — An Example Of An Indigenous Elementary School In Zenai Township, Nanto County. Master Thesis of Providence University.

[6] Horning, K. T. (1997). From cover to evaluating and reviewing children's books. New York: Harpercollins publishers.

[7] Malo, E., & Bullard, J. (2000). Storytelling and the emergent reader (Report No. CS 217 352). Auckland, New Zealand: The 18th International Reading Association World Congress on Reading (ERIC Document Reproduction Service No. ED448464).

[8] Isbell, R., Sobol, J., Lindauer, L., & Lowrance, A. (2004). The effects of storytelling and story reading on the oral language complexity and story comprehension of young children. Early Childhood Education Journal, 32(3), 157-163.

[9] Johnson, P. (1992). Human-computer interaction: Psychology, task analysis and software engineering. London: McGraw-Hill.

[10] Tennyson, R. D., & Park, S. I. (1987). Artificial intelligence and computer-based learning. In R. M. Gagne (Ed.), Instructional technology: Foundations (pp. 319-342). Hillsdale, New York: Lawrence Erlbaum Associates.

[11] Argyris, C., & Schön, D. (1974). Theory in practice: Increasing professional effectiveness. London: Jossey- Bass.

[12] Hitchcock, G., & Hughes, D. (1994). Research and the teacher: A qualitative introduction to school-based research (2nd ed.). New York, NY: Routledge.

Almost Boolean fuzzy rings

B A Alaba and Y G Wondifraw

Department of Mathematics, College of Science, Bahir Dar University, Bahir Dar, Ethiopia

berhanu_assaye@yahoo.com, yohannesg27@gmail.com

Abstract. In this paper we study the class of relatively complemented Almost Distributive Fuzzy Lattices in detail. We introduce the concept of an Almost Boolean Fuzzy Rings as a generalization of a Boolean Fuzzy Rings. We also establish a one-to-one correspondence between relatively complemented Almost Distributive Fuzzy Lattices and the class of an Almost Boolean Fuzzy Rings.

1. Introduction

The concept of an Almost Distributive Lattice (ADL) was introduced by Swamy U M and Rao G C [1] as a common abstraction of almost all the existing ring theoretic generalizations of a Boolean algebra on one hand and the class of distributive lattices on the other. Again, the concept of relatively complemented ADLs and an Almost Boolean Rings are introduced by Rao G C [2]. A relatively complemented ADL is an ADL in which every interval is a complemented lattice. An Almost Boolean Ring is a triple (R, +, ·, 0) satisfying all the properties of a Boolean ring except possibly the associativity of +. On the other hand the concept of a fuzzy set was first introduced by Zadeh L A [3] and this concept was adapted by Goguen J A [4] and Sanchez E [5] to define and study fuzzy relations. In 1994, Ajmal N and Thomas K V [6] defined a fuzzy lattice as a fuzzy algebra and characterized fuzzy sublattices. In 2009, Chon I [7], considering the notion of fuzzy order of Zadeh L A [8], introduced a new notion of fuzzy lattice and studied the level sets of fuzzy lattices. In this paper we introduce a new mathematical notion, relatively complemented ADFLs and an Almost Boolean Fuzzy Rings and characterized some properties of them using the fuzzy partial order relation and fuzzy lattice defined by Chon I.

2. Preliminary

First, we recall certain definitions and properties of ADLs, Compatibility and Fuzzy Partial Order Relations that are required in the paper.

Definition 2.0.1. [1] An algebra $(R, \vee, \wedge, 0)$ of type (2, 2, 0) is called an Almost Distributive Lattice, abbreviated as ADL, if it satisfies the following axioms:

1. $a \vee 0 = 0$
2. $0 \wedge a = 0$
3. $(a \vee b) \wedge c = (a \wedge c) \vee (b \wedge c)$
4. $a \wedge (b \vee c) = (a \wedge b) \vee (a \wedge c)$
5. $a \vee (b \wedge c) = (a \vee b) \wedge (a \vee c)$
6. $(a \vee b) \wedge b = b$

for all $a, b, c \in R$.

IOP Conf. Series: Journal of Physics: Conf. Series **814** (2017) 012009 doi:10.1088/1742-6596/814/1/012009

Definition 2.0.2. [1] For any a, b ∈ R, we say that a is less than or equal to b and write $a \leq b$ if $a \wedge b = a$ or equivalently, $a \vee b = b$.

Lemma 2.0.3. [1] For any a, b, c ∈ R,

1. $a \wedge a = a$
2. $a \wedge (a \vee b) = a = a \vee (a \wedge b)$
3. $(a \vee b) \wedge a = a$
4. $a \wedge b \wedge c = b \wedge a \wedge c$
5. $(a \vee b) \wedge c = (b \vee a) \wedge c$
6. $\wedge$ is associative in R .

Lemma 2.0.4. [1] If $a_1, a_2, \ldots, a_n, b \in R$ and $(i_1, i_2, \ldots, i_n)$ is any permutation of (1, 2, …., n), then $(a_{i1} \vee a_{i2} \vee \ldots \vee a_{in}) \wedge b = (a_1 \vee a_2 \vee \ldots \vee a_n) \wedge b$.

Corollary 2.0.5.[2] For any a, b ∈ R, $a \vee b = b \vee a$ whenever $a \wedge b = 0$.

Corollary 2.0.6. [2] For any a, b, c, d ∈ R, we have $\{a \vee (b \vee c)\} \wedge d = \{(a \vee b) \vee c\} \wedge d$.

Remark: If $a \leq b$ and x ∈ R, then $x \vee a \leq x \vee b$.

Definition 2.0.7. [2] Let (L, ∨, ∧) be an ADL. By interval in L we mean the set $[a, b] = \{x \in R : a \leq x \leq b\}$ for some a, b ∈ L such that $a \leq b$.

Note: A Boolean algebra is a bounded, distributive complemented lattice.

Definition 2.0.8. [2] For any a, b ∈ R a is said to be compatible with b (written $a \sim b$) if $a \wedge b = b \wedge a$ or equivalently, $a \vee b = b \vee a$.

Lemma 2.0.9. [2] Let L be a relatively complemented ADL with 0. Then for any a, b, c ∈ L, we have the following:

1. $a^a = 0 = a^0$ and $0^a = a$
2. $a^c = (a \wedge c)^c$
3. $(a \wedge c)^{b \wedge c} = a^b \wedge c$
4. $(a \vee b)^c = a^c \wedge b^c$
5. $a \wedge b^c = (a \wedge b)^{a \wedge c}$

Definition 2.0.10. [2] An algebra (R, +, ·, 0) of type (2, 2, 0) is called an Almost Boolean Ring if it satisfies the following axioms:

R1. $x + 0 = x$
R2. $x + x = 0$
R3. $x^2 = x$
R4. $(x y)z = x(y z)$
R5. $x(y + z) = x y + x z$
R6. $(x + y) z = x z + y z$
R7. $\{(x + y) + z\} t = \{x + (y + z)\} t$
for all x, y, z, t ∈ R.

Lemma 2.0.11. [2] For any a ∈ R, we have the following:

1. $a \cdot 0 = 0 \cdot a = a$
2. The set $Ra = \{xa : x \in R\}$ is either {0} or a group of characteristic two under the induced operations + and · .

Corollary 2.0.12. [2] For any a, b, c ∈ R, $(a + b) c = (b + a) c$.

Corollary 2.0.13. [2] For any a, b, c ∈ R, abc = bac.

As a particular case of the above corollary we have,

Corollary 2.0.14. [2] For any a, b ∈ R, bab = ab.

Theorem 2.0.15. [2] For any a, b ∈ R, $a + b = b + a$.

Definition 2.0.16.[7] Let X be a set. A function $A : X \times X \longrightarrow [0, 1]$ is called a fuzzy relation in X. The fuzzy relation A in X is reflexive iff A(x, x) = 1 for all x ∈ X, A is transitive iff $A(x, z) \geq sup_{y \in X} min(A(x, y), A(y, z))$, and A is antisymmetric iff A(x, y) > 0 and A(y, x) > 0 implies x = y. A fuzzy relation is fuzzy partial order relation if A is reflexive, antisymmetric and transitive. A fuzzy partial order relation A is a fuzzy total order relation iff A(x, y) > 0 or A(y, x) > 0 for all x, y ∈ R. If A is fuzzy partial order relation in a set X, then (X, A) is called a fuzzy partially ordered set or a fuzzy poset. If B is a fuzzy total order relation in a set X, then (X, B) is called a fuzzy totally ordered set or a fuzzy chain.

3. Almost Boolean fuzzy rings (ABFRs)

In this section we introduce a new mathematical notion an Almost Distributive Fuzzy Lattice, relatively complemented ADFL and an Almost Boolean Fuzzy Rings and we investigate and prove some results.

3.1 Relatively complemented ADFL

By interval in an ADFL we mean an interval in an ADL.

Definition 3.1.1. Let (R, ∨, ∧, 0) be an algebra oftype (2, 2, 0) and (R, A) be a fuzzy poset. Then we call (R, A) is an Almost Distributive Fuzzy Lattice (ADFL) if the following axioms are satisfied:

1. $A(a, a \vee 0) = A(a \vee 0, a) = 1$,
2. $A(0, 0 \wedge a) = A(0 \wedge a, 0) = 1$,
3. $A((a \vee b) \wedge c, (a \wedge c) \vee (b \wedge c)) = A((a \wedge c) \vee (b \wedge c), (a \vee b) \wedge c) = 1$,
4. $A(a \wedge (b \vee c), (a \wedge b) \vee (a \wedge c)) = A((a \wedge b) \vee (a \wedge c), a \wedge (b \vee c)) = 1$,
5. $A(a \vee (b \wedge c), (a \vee b) \wedge (a \vee c)) = A((a \vee b) \wedge (a \vee c), a \vee (b \wedge c)) = 1$,
6. $A((a \vee b) \wedge b, b) = A(b, (a \vee b) \wedge b) = 1$,

for all a, b, c ∈ R.

Theorem 3.1.2. Let (R, A) be an ADFL. Then $a = b \Leftrightarrow A(a, b) = A(b, a) = 1$.

Definition 3.1.3. Let (R, A) be an ADFL. Then for any a, b∈ R, $a \leq b$ if and only if $A(a, b) > 0$.

In view of the above definition, we have the following theorem.

Theorem 3.1.4. If (R, A) is an ADFL then $a \wedge b = a$ if and only if $A(a, b) > 0$.

Lemma 3.1.5. Let (R, A) be an ADFL and a, b ∈ R such that $a \neq b$. If $A(a, b) > 0$ then $A(b, a) = 0$.

Lemma 3.1.6. Let (R, A) be an ADFL. For any a, b ∈ R, $A(a \wedge b, 0) > 0$ if and only if $A(b \wedge a, 0) > 0$.

Definition 3.1.7. An ADFL (R, A) is said to be relatively complemented if every interval in R is a Boolean algebra.

Lemma 3.1.8. An ADFL (R, A) is relatively complemented if and only if for any a, b ∈ R , there exists x ∈ R , such that $A(a \vee x, a \vee b) = A(a \vee b, a \vee x) = 1$ and $A(a \wedge x, 0) > 0$, in this case, x is unique which we denote by a^b.

Proof: (⇒) An ADFL (R, A) is relatively complemented if every interval in R is a Boolean algebra. Hence [0, a] is a Boolean algebra, which implies [0, a] is complemented lattice. Let b ∈ R such that $A(a, b) > 0$. Then interval $[0, a \vee b]$ is complemented and $a \in [0, a \vee b]$. If x is the complement of a in $[0, a \vee b]$, then $A(a \vee x, a \vee b) = A(a \vee b, a \vee x) = 1$ and $A(a \wedge x, 0) > 0$. Let $y \in [0, a \vee b]$ satisfying $A(a \vee y, a \vee b) = A(a \vee b, a \vee y) = 1$ and $A(a \wedge y, 0) > 0$.

Claim: $A(x, y) = A(y, x) = 1$.

Since $A(x, y \wedge x) = 1$ and $A(x \wedge y, y) = 1$ and both $x, y \in [0, a \vee b]$ and $[0, a \vee b]$ is a Boolean algebra then $x \wedge y = y \wedge x$ and hence $A(y \wedge x, y) = 1$. Hence by using the transitive property of A, $A(x, y) = A(y, x) = 1$.Therefore the complement is unique.

(⇐) Suppose for any a, b ∈ R, there exists a unique x in R such that $A(a \vee x, a \vee b) = A(a \vee b, a \vee x) = 1$ and $A(a \wedge x, 0) > 0$.

Claim: The interval [a, b] in R is complemented lattice.

Now, let a, b ∈ R such that $A(a, b) > 0$ and $x \in [a, b]$ then by hypothesis there exists a unique y ∈ R, such that $A(x \vee y, b) = A(b, x \vee y) = 1$ and $A(x \wedge y, 0) > 0$ since $x \leq b$ then $x \vee y = x \vee b = b$. Since from ADL $y \wedge (y \vee x) = y$ then, $y \leq y \vee x = x \vee y = b$. Now it is easy to verify that $a \vee y \in [a, b]$ and it is the complement of x in [a, b]. Therefore the interval [a, b] in R is a Boolean algebra and hence an ADFL (R, A) is relatively complemented.

Note: Let (R, A) be a relatively complemented ADFL for any a, b ∈ R, we write a^b for the unique element x ∈ R with the property $A(a \wedge a^b, 0) > 0$ and $A(a \vee a^b, a \vee b) = A(a \vee b, a \vee a^b) = 1$.

Lemma 3.1.9. If an ADFL (R, A) is relatively complemented and a, b ∈ R, then $A(a^b, b) > 0$.

Proof: Suppose an ADFL (R, A) is relatively complemented and a, b ∈ R.

Now, $A(a^b \wedge b, a^b) = A(0 \vee (a^b \wedge b), a^b) = A((a^b \wedge a) \vee (a^b \wedge b), a^b) = A(a^b \wedge (a \vee b), a^b) = A(a^b \wedge (a \vee b), a^b) = A(a^b \wedge (a^b \vee a), a^b) = A(a^b, a^b) = 1$. Similarly, $A(a^b, a^b \wedge b) = 1$.
Hence $A(a^b \wedge b, a^b) = A(a^b, a^b \wedge b) = 1$. Then we have $a^b \leq b$.Therefore $A(a^b, b) > 0$.

Lemma 3.1.10. If an ADFL (R, A) is relatively complemented and $a, b \in R$, then $A(a^b \vee b^a, b^a \vee a^b) = 1$.

Proof: Suppose (R, A) is an ADFL and $a, b \in R$.
$A(a^b \wedge b^a, 0) = A(a^b \wedge b^a \wedge a, 0) = A(b^a \wedge a^b \wedge a, 0) = A(b^a \wedge 0, 0) = A(0, 0) = 1$.
Similarly, $A(0, a^b \wedge b^a) = 1$.Hence we have $a^b \wedge b^a = 0$.Therefore, $A(a^b \vee b^a, b^a \vee a^b) = 1$.

Lemma 3.1.11. If an ADFL (R, A) is relatively complemented. For any $a, b, c \in R$, then
1. $A(a^b \wedge b^c, b^c \wedge a^b) = 1$.
2. $A(b^c \wedge c, c \wedge b^c) = 1$.

Proof: Direct using Lemma 3.1.9.

Lemma 3.1.12. Let (R, A) be an ADFL. For any $a, b, c \in R$ such that $a \wedge b = 0 = a \wedge c$. Then $A(a \vee (b \vee c), (a \vee b) \vee c) = A((a \vee b) \vee c, a \vee (b \vee c)) = 1$.

Lemma 3.1.13. If an ADFL (R, A) is relatively complemented and $a, b, c \in R$. Then we have the following:
1. $A((a \wedge c)^c, a^c) = 1$
2. $A((a \vee b)^c, a^c \wedge b^c) = 1$
3. $A((a \wedge b)^c, a^c \vee b^c) = 1$
4. $A((a \wedge b)^{a \wedge c}, a \wedge b^c) = 1$
5. $A((b \wedge a)^{c \wedge a}, b^c \wedge a) = 1$.

Proof: Let (R, A) be relatively complemented ADFL and $a, b, c \in R$
1. Since $A(a \wedge c \wedge (a \wedge c)^c, 0) > 0$ and $A((a \wedge c) \vee (a \wedge c)^c, (a \wedge c) \vee c) = A((a \wedge c) \vee c, (a \wedge c) \vee (a \wedge c)^c) = 1$. Hence $(a \wedge c)^c$ is the complement of $a \wedge c$.
Now, $A(a \wedge c \wedge a^c, 0) = A(c \wedge a \wedge a^c, 0) = A(c \wedge 0, 0) = A(0, 0) = 1$. Hence $A(a \wedge c \wedge a^c, 0) > 0$.
$A((a \wedge c) \vee a^c, c) = A(a^c \vee (a \wedge c), c) = A((a^c \vee a) \wedge (a^c \vee c), c) = A((a \vee a^c) \wedge (a^c \vee c), c) = A((a \vee c) \wedge (a^c \vee c), c) = A((a \vee c) \wedge c, c) = A(c, c) = 1$. Similarly, $A(c, (a \wedge c) \vee a^c) = 1$.
Therefore $A((a \wedge c) \vee a^c, c) = A(c, (a \wedge c) \vee a^c) = 1$. Hence a^c is the complement of $a \wedge c$.
Hence by uniqueness of complement $A((a \wedge c)^c, a^c) = 1$.
We can prove others similarly.
By using Lemma 3.1.12 the proof of the following theorem is direct.

Theorem 3.1.14. Let (R, A) be relatively complemented ADFL. Then for any $a, b, c \in R$, we have $A((a \vee b) \vee c, a \vee (b \vee c)) = 1$.
By using the concept of uniqueness of complement, the following Lemma is direct.

Lemma 3.1.15. Let (R, A) be relatively complemented ADFL. For any $a, b, c \in R$,
1. $A(a^a, 0) = 1$
2. $A(0, a^0) = 1$
3. $A((a \wedge b)^c, a^c \vee b^c) = 1$
4. $A(a^a, a^0) = 1$
5. $A(0^a, a) = 1$
6. $A(b^{a \wedge c}, a \wedge b^c) = 1$
7. $A(a^c, (a \wedge c)^c) = 1$.

3.2 Almost Boolean Fuzzy Rings

Definition 3.2.1. Let $(R, +, \cdot, 0)$ be an algebra of type (2, 2, 0) and (R, A) be a fuzzy poset. Then (R, A) is a fuzzy ring if it satisfies the following axioms:
1. $A((a + b) + c, a + (b + c)) = A(a + (b + c), (a + b) + c) = 1$
2. $A(a + 0, a) = A(a, a + 0) = 1$
3. $A(a + (-a), 0) = A(0, a + (-a)) = 1$
4. $A(a + b, b + a) = A(b + a, a + b) = 1$
5. $A((ab)c, a(bc)) = A(a(bc), (ab)c) = 1$
6. $A(a(b + c), ab + ac) = A(ab + ac, a(b + c)) = 1$

IOP Conf. Series: Journal of Physics: Conf. Series **814** (2017) 012009 doi:10.1088/1742-6596/814/1/012009

7. $A((b + c)a, ba + ca) = A(ba + ca, (b + c)a) = 1$
for all $a, b, c, -a \in R$.

Definition 3.2.2. Let(R, A) be a fuzzy ring. Then (R, A) is a Boolean fuzzy ring if and only if $A(a^2, a) = A(a, a^2) = 1$, for all $a \in R$

Definition 3.2.3. Let $(R, +, \cdot, 0)$ be an algebra of type (2, 2, 0) and (R, A) be a fuzzy poset. Then we call (R, A) is an Almost Boolean Fuzzy Ring abbreviated as ABFR if the following axioms are satisfied:

RF1. $A(a + 0, a) = A(a, a + 0) = 1$
RF2. $A(a + a, 0) = A(0, a + a) = 1$
RF3. $A((ab)c, a(bc)) = A(a(bc), (ab)c) = 1$
RF4. $A(a^2, a) = A(a,a^2) = 1$
RF5. $A(a(b + c), ab + ac) = A(ab + ac, a(b + c)) = 1$
RF6. $A((a + b)c, ac + bc) = A(ac + bc, (a + b)c) = 1$
RF7. $A(\{a + (b + c)\}d, \{(a + b)+ c\}d) = A(\{(a + b)+ c\}d, \{a + (b + c)\}d) = 1$
for all $a, b, c, d \in R$.

It is possible to show that the above seven properties in an ABFR are independent with examples. From the definition of ABR and ABFR the following theorem is direct.

Theorem 3.2.4. Let (R, A) be a fuzzy poset. Then R is an ABR if and only if (R, A) is an ABFR.

Lemma 3.2.5. Let (R, A) be an ABFR. For any $a, b, c \in R$, we have
1. $A(a0, 0) = 1$
2. $A(0, a0) = 1$
3. $A(a0, 0a) = 1$

Proof: Suppose (R, A) is an ABFR and $a, b, c \in R$
1. $A(a0, 0) = A(a(a + a), 0) = A((a^2+ a^2), 0) = A(a + a, 0) = A(0, 0) = 1$.
Hence $A(a0, 0) = 1$.
2. $A(0, 0a) = A(0, (a + a)a) = A(0, a^2+ a^2) = A(0, a + a) = A(0, 0)= 1$.
Hence $A(0a, 0) = 1$.
3. $A(a0, 0a) \geq sup_{c\in R} min(A(a0, c), A(c, 0a))$
$\geq min(A(a0, 0), A(0, 0a)) = min(1, 1) = 1$.
Hence $A(a0, 0a) = 1$.

Theorem 3.2.6. Let (R, A) be a relatively complemented ADFL. Define a binary operation + on R by $a + b = a^b \vee b^a$. Then (R, A) is an ABFR.

Proof: Suppose (R, A) is relatively complemented ADFL.

Claim: (R, A) is an ABFR, where $a + b =a^b \vee b^a$.
1. $A(a + 0, a) = A(a^0\vee 0^a, a) = A(0\vee a, a) = A(a, a) = 1$.
Similarly, $A(a, a + 0) = 1$. Therefore $A(a + 0, a) = A(a, a + 0) = 1$.
2. $A(a + a, 0) = A(a^a\vee a^a, 0) = A(0\vee 0, 0) = A(0, 0)= 1$
Similarly, $A(0, a + a) = 1$.Therefore $A(a + a, 0) = A(0, a + a) = 1$.
3. $A((a\wedge b) \wedge c, a\wedge (b\wedge c)) = A(a \wedge(b\wedge c), a \wedge (b\wedge c)) = 1$
Similarly, $A(a \wedge(b \wedge c), (a \wedge b)\wedge c) = 1$.
Therefore $A((a \wedge b)\wedge c, a\wedge(b \wedge c)) = A(a \wedge(b \wedge c), (a \wedge b)\wedge c) = 1$
4. $A(a \wedge a, a) = A(a, a) = 1 = A(a, a \wedge a)$
5. $A(a \wedge(b + c), (a\wedge b) + (a\wedge c)) = A(a\wedge(b^c\vee c^b), (a\wedge b) + (a\wedge c)) = A((a\wedge b^c)\vee(a\wedge c^b), (a\wedge b) + (a\wedge c))=$ $A((a \wedge b)^{a\wedge c} \vee (a \wedge c)^{a\wedge b}, (a\wedge b) + (a\wedge c)) = A((a\wedge b) + (a\wedge c), (a\wedge b) + (a\wedge c))= 1$.
Similarly, $A((a \wedge b) + (a \wedge c), a \wedge (b + c))= 1$.Therefore $A(a \wedge(b + c), (a \wedge b) + (a \wedge c)) = A((a \wedge b) + (a \wedge c), a \wedge (b + c))= 1$.
6.$A((a + b)\wedge c, (a\wedge c) + (b\wedge c))= A((a^b\vee b^a)\wedge c, (a\wedge c) + (b\wedge c))=A((a^b\wedge c) \vee (b^a\wedge c), (a\wedge c) + (b\wedge c))=$ $A((a \wedge c)^{b\wedge c} \vee (b \wedge c)^{a\wedge c}, (a\wedge c) + (b\wedge c))= A((a \wedge c) + (b \wedge c), (a \wedge c) + (b \wedge c))= 1$.
Similarly, $A((a \wedge c) + (b \wedge c), (a + b) \wedge c) = 1$.
Therefore $A((a + b) \wedge c, (a \wedge c) + (b \wedge c)) = A((a \wedge c) + (b \wedge c), (a + b) \wedge c) = 1$.
7. $A(\{a + (b + c)\}\wedge d, \{(a + b) + c\}\wedge d) = A(\{a + (b^c\vee c^b)\}\wedge d, \{(a + b) + c\}\wedge d)$
$= A(\{a^{(b^c\vee c^b)} \vee(b^c\vee c^b)^a\}\wedge d,\{(a + b) + c\}\wedge d) = A(\{ a\vee b^c\vee c^b\vee b^c\vee c^b\vee a \}\wedge d, \{(a + b) + c\}\wedge d)$

$= A(\{a^b \vee b^a \vee c \vee c \vee a^b \vee b^a\} \wedge d, \{(a+b)+c\} \wedge d) = A(\{(a^b \vee b^a)^c \vee c^{a^b \vee b^a}\} \wedge d, \{(a+b)+c\} \wedge d)$
$= A(\{(a+b)^c \vee c^{a+b}\} \wedge d, \{(a+b)+c\} \wedge d) = A(\{(a+b)+c\} \wedge d, \{(a+b)+c\} \wedge d) = 1.$
Similarly, $A(\{(a+b)+c\} \wedge d, \{a+(b+c\} \wedge d) = 1$
Therefore $A(\{a+(b+c)\} \wedge d, \{(a+b)+c\} \wedge d) = A(\{(a+b)+c\} \wedge d, \{a+(b+c\} \wedge d) = 1.$
Hence (R, A) is an Almost Boolean Fuzzy Rings.
Now, we finish this section by proving the converse of the above theorem.
Theorem 3.2.7. Let (R, A) be an ABFR. Define the operation $\vee$ on R by $a \vee b = a + (b + ab)$. Then (R, A) is relatively complemented ADFL.
Proof: Let (R, A) be an ABFR and $a \vee b = a + (b + ab)$.
Claim: (R, A) is relatively complemented ADFL
First we need to show (R, A) is an ADFL.
1. $A(a \vee 0, a) = A(a + (0 + a0), a) = A(a + (0 + 0), a) = A(a, a) = 1$
Similarly, $A(a, a \vee 0) = 1$. Therefore $A(a \vee 0, a) = A(a, a \vee 0) = 1$.
2. $A(0a, 0) = A(0, 0) = 1$
Similarly, $A(0, 0a) = 1$. Therefore $A(0a, 0) = A(0, 0a) = 1$.
3. $A((a \vee b)c, ac \vee bc) = A((a + (b + ab))c, ac \vee bc) = A(ac + (b + ab)c, ac \vee bc) = A(ac + (bc + abc), ac \vee bc) = A(ac + (bc + acbc), ac \vee bc) = A(ac \vee bc, ac \vee bc) = 1$.
Similarly $A(ac \vee bc, (a \vee b)c) = 1$.
Therefore $A((a \vee b)c, ac \vee bc) = A(ac \vee bc, (a \vee b)c) = 1$.
4. $A(a(b \vee c), ab \vee ac) = A(a(b + (c + bc)), ab \vee ac) = A(ab + a(c + bc), ab \vee ac) = A(ab + (ac + abc), ab \vee ac) = A(ab + (ac + abac), ab \vee ac) = A(ab \vee ac, ab \vee ac) = 1$
Similarly, $A(ab \vee ac, a(b \vee c)) = 1$. Therefore $A(a(b \vee c), ab \vee ac) = A(ab \vee ac, a(b \vee c)) = 1$.
5. $A(a \vee (bc), (a \vee b)(a \vee c)) = A(a \vee (bc), (a + b + ab)(a + c + ac)) = A(a \vee (bc), a(a + c + ac) + b(a + c + ac) + ab(a + c + ac)) = A(a \vee (bc), a^2 + ac + a^2c + ba + bc + bac + aba + abc + abac) = A(a \vee (bc), a + ac + ac + ba + bc + bac + aba + abc + abac) = A(a \vee (bc), a + bc + abc) = A(a \vee (bc), a \vee (bc)) = 1$.
Similarly, $A((a \vee b)(a \vee c), a \vee (bc)) = 1$. Therefore $A(a \vee (bc), (a \vee b)(a \vee c)) = A((a \vee b)(a \vee c), a \vee (bc)) = 1$
6. $A((a \vee b)b, b) = A((a + b + ab)b, b) = A(ab + b^2 + ab^2, b) = A(ab + b + ab, b) = A(b, b) = 1$
Similarly, $A(b, (a \vee b)b) = 1$. Therefore $A((a \vee b)b, b) = A(b, (a \vee b)b) = 1$.
Hence (R, A) is an ADFL.
To show (R, A) is relatively complemented ADFL
Let $a, b \in R$, then
$A(a(b + ab), 0) = A(ab + a^2b, 0) = A(ab + ab, 0) = A(0, 0) = 1 > 0$ and
$A(a \vee (b + ab), a \vee b) = A(a + ((b + ab) + a(b + ab)), a \vee b) = A(a + (b + ab + ab + a^2b), a \vee b) = A(a + (b + ab + ab + ab), a \vee b) = A(a \vee b, a \vee b) = 1$
Similarly, $A(a \vee b, a \vee (b + ab)) = 1$. Which implies $A(a \vee (b + ab), a \vee b) = A(a \vee b, a \vee (b + ab)) = 1$.
Therefore by Lemma 3.1.8 (R, A) is a relatively complemented ADFL in which for any $a, b \in R$,
$a^b = b + ab$.

4. References

[1] Swamy U M and Rao G C 1981 Almost Distributive Lattices *Jour. Aust. Math. Soc.* **31** pp77-91
[2] Rao G C 1980 Almost Distributive Lattice Doctoral Thesis Department of Mathematics Andhra University
[3] Zadeh L A 1965 Fuzzy Sets *Inform. and Control* **8** pp 338-53
[4] Goguen J A 1967 L – fuzzy sets *J. Math. Anal. Appl.* **18** pp 145-74
[5] Sanchez E 1976 Resolution of composite fuzzy relation equation *Inform. and con.* **30** pp 38- 48
[6] Ajmal N and Thomas K V 1994 Fuzzy Lattices *Info. Sci.* **79** pp 271-91
[7] Chon I 2009 Fuzzy Partial Order Relations and Fuzzy Lattices *Kor. J. Math.* **17** pp 361-74
[8] Zadeh L A 1971 Similarity Relations and Fuzzy Orderings *Info. Sci.* **3** pp 177-200

IOP Conf. Series: Journal of Physics: Conf. Series **814** (2017) 012010 doi:10.1088/1742-6596/814/1/012010

Efficient large scale commute time embedding

H I Hahn

Department of Information and Communications Eng., Hankuk University of Foreign Studies, 89 Wangsan, Mohyun, Yongin , Kyonggi-Do, 449-791, Korea

hihahn@hufs.ac.kr

Abstract. Commute time embedding involves computing eigenfunctions of the graph Laplacian matrix. Spectral decomposition requires computational burden proportional to $O(n^3)$, which may not be suitable for large scale dataset. This paper proposes computationally efficient commute time embedding by applying Nyström method to the normalized graph Laplacian. The performance of the proposed algorithms is analysed by checking the embedding results on a patch graph.

1. Introduction

When dataset lies on a linear subspace, PCA(principal component analysis) is most useful and optimal for embedding as well as dimensionality reduction in terms of maintaining maximum variance of dataset. However, when dataset lies on a nonlinear space, PCA introduces severe error. The manifold learning algorithms replace PCA on a nonlinear space. Although there are lots of manifold learning algorithms, commute time embedding is known to be most suitable for dimensionality reduction.

Basically, manifold learning algorithms usually use kernel-based methods to represent the data of nonlinear structure, because it is more clear in the feature space rather than in the original input space. In kernel-based methods, kernel matrix is defined when input data are mapped to the feature space using a map $\Phi : \mathcal{X} \rightarrow \mathcal{F}$. Kernel matrix $G \in \mathbf{R}^{n \times n}$, whose element $G_{ij} = k(x_i, x_j) = \langle \Phi(x_i), \Phi(x_j) \rangle$, can be represented in terms of an inner product between feature vectors. For example, ISOMAP [1] computes the matrix whose elements are the geodesic distance between pairs of nodes. On the contrary, Laplacian eigenmap [2] and commute time embedding [3] employ graph Laplacian matrices as kernels. This scheme needs to compute the spectral decomposition of the kernel matrix, whose computational burden increases in the order of $O(n^3)$, where n is the size of dataset. It is hardly applicable when n is sufficiently large. Lots of researchers have tried to solve this problem by accelerating the computational time of spectral decomposition. Nyström method, proposed by Williams, et al. [4], is adopted in [5] to compute the kernel eigenfunctions approximately from the affinity matrix composed of randomly chosen samples. In spite of lots of researches on approximation of kernel matrices, there have been rarely known about approximation of the normalized graph Laplacian. The properties that Graph Laplacian matrix is symmetric and positive semidefinite (SPSD) should be preserved even after its approximation. However, the method proposed by Choromanska, et al. [6] does not satisfy it any more. They try to create a submatrix by random sampling the columns of graph Laplacian and normalizing it directly. However, it induces a severe error because the submatrix is no longer SPSD. This paper proposes a new method to reduce the approximation error significantly. Commute time embedding is performed on the patch graph, so that each patch is mapped into the corresponding vector.

2. Review of Commute Time Embedding

2.1. Construction of patch graph

Suppose that maximally overlapped patches of size p samples are extracted around each time sample in the following way:

$$\hat{\bar{\mathbf{x}}}_n = \frac{\bar{\mathbf{x}}_n}{\|\bar{\mathbf{x}}_n\|} \in S^{p-2}, \quad n = 1, 2, \cdots, N, \tag{1}$$

where $\bar{\mathbf{x}}_n$ is the mean centered version of $\mathbf{x}_n = (x[n], x[n+1], \cdots, x[n+p-1])^T \in \mathbf{R}^p$. The patch $\hat{\bar{\mathbf{x}}}_n$ is obtained by normalizing with the magnitude. The weight along the edge connecting the nodes v_i with v_j, which are associated with $\hat{\bar{\mathbf{x}}}_i$ and $\hat{\bar{\mathbf{x}}}_j$, respectively, is defined as follows:

$$w(i,j) = \begin{cases} \exp\left(-\left\|\hat{\bar{\mathbf{x}}}_i - \hat{\bar{\mathbf{x}}}_j\right\|^2 / 2\sigma^2\right) & v_i, v_j : \text{connected} \\ 0 & \text{otherwise.} \end{cases} \tag{2}$$

Given a set of patches $\hat{\bar{\mathbf{x}}}_1, \cdots, \hat{\bar{\mathbf{x}}}_N$ and $w(i,j)$, we can construct a graph, where two vertices v_i and v_j are connected with a weight $w(i,j)$.

2.2. Commute time embedding

Given an affinity matrix $W \in \mathbf{R}^{N\times N}$, whose entries are $w(u,v)$, the degree matrix D is computed to be diagonal with entries $d_u = \sum_{v=1}^{N} w(u,v)$ and the graph Laplacian is defined as $L = D - W$. Let $L = U\Lambda U^T$ be the spectral decomposition of L, where U comprises all eigenvectors and Λ is diagonal with the corresponding eigenvalues $\lambda_1, \cdots, \lambda_N$. The Moore-Penrose inverse of L, $L^\dagger = U\Lambda^\dagger U^T$ is defined as

$$\lambda_i^\dagger = \begin{cases} 1/\lambda_i & \lambda_i \neq 0 \\ 0 & \textit{otherwise.} \end{cases} \tag{3}$$

where $\Lambda^\dagger = diag(\lambda_1^\dagger, \cdots, \lambda_N^\dagger)$. Then, the commute time between two vertices v_i and v_j is given as $c(i,j) = vol(e_i - e_j)^T L^\dagger (e_i - e_j)$, where $vol = \sum_{u,v} w(u,v)$ and $e_i = [0, \cdots, 0, 1, 0, \cdots, 0]^T$ is defined as the ith column of the identity matrix I. Hence, $\sqrt{c(i,j)}$ can be considered a Mahalanobis distance with a weighting matrix $vol \cdot L^\dagger$, and $c(i,j)$ can be rephrased as

$$c(i,j) = vol \cdot (e_i - e_j)^T U\Lambda^\dagger U^T (e_i - e_j) = (z_i - z_j)^T (z_i - z_j), \tag{4}$$

where $z_i = \sqrt{vol}\left(u_{i,2}/\sqrt{\lambda_2}, \cdots, u_{i,N}/\sqrt{\lambda_N}\right)$. Here $u_{j,k}$ is the jth element of u_k. L can be normalized as $L_{sym} = D^{-1/2} L D^{-1/2}$. If we use L_{sym} instead of L, z_i can be rephrased as:

$$z_i = \sqrt{vol}\left(\frac{v_{i,2}}{\sqrt{\lambda_2' d_i}}, \cdots, \frac{v_{i,N}}{\sqrt{\lambda_N' d_i}}\right), \tag{5}$$

where λ_k' is the eigenvalue of L_{sym} and v_k is the corresponding eigenvector. For the dimensionality reduction, we can use only the first q components corresponding to the lower eigenvectors. λ_k''s are assumed to be sorted in the following way [3]:

$$0 \le \lambda_1' \le \lambda_2' \le \cdots \lambda_N' < 2. \tag{6}$$

3. Approximate commute time embedding using Nyström method

Let C be the $N \times c$ matrix obtained by random sampling c columns of an adjacency matrix W uniformly without replacement and A be the $c \times c$ matrix obtained by extracting the rows of the same indices as the sampled c columns of W. The rows and columns of W are rearranged as follows:

$$W = \begin{bmatrix} A & B^T \\ B & H \end{bmatrix}, \qquad C = \begin{bmatrix} A \\ B \end{bmatrix}. \tag{7}$$

Since W is SPSD, A is also SPSD. Choromanska, et al. [6] approximate $L_{sym} = D^{-1/2} L D^{-1/2}$ by computing $\tilde{L}_{sym} \leftarrow \hat{I} - \sqrt{c/N} D^{-1/2} \cdot C \cdot \Delta^{-1/2}$. Here, $\hat{I}$ is obtained by sampling the columns of the identity matrix I where Ω represents the index set $\{i_1, i_1, \cdots, i_c\}$ sampled uniformly, which is denoted as $\hat{I} \leftarrow I(\ :\ , \Omega\ )$. And the degree matrices $D \in \mathbf{R}^{N \times N}$, $\Delta \in \mathbf{R}^{c \times c}$ are defined as follows:

$$D(i,i) = \sum_{j=1}^{c} C(i,i), \qquad \Delta(i,i) = \sum_{j=1}^{N} C(j,i). \tag{8}$$

They approximate the eigenvalues and eigenvectors of L_{sym} using the spectral decomposition of $\hat{L}_{sym} \leftarrow \tilde{L}_{sym}(\ \Omega, :\ )$. The eigenvalues and eigenvectors of $\hat{L}_{sym}$ deviate from those of L_{sym} and do not even satisfy the properties that the least eigenvalue be zero, since it is no longer SPSD.

In this paper, we propose a different method to approximate L_{sym}, which preserves the properties that L_{sym} be SPSD even though it is approximated. At first, A is normalized using its degree matrix D_A to get $\hat{A} \leftarrow D_A^{-1/2} A D_A^{-1/2}$, where $D_A(i,i) := \sum_j A(i,j)$. However, B should be normalized differently, as given in the fourth step of the algorithm explained in figure 1, since B is neither a symmetric nor a square matrix. Finally, we approximate the eigenvalues and eigenvectors of L_{sym} via the spectral decomposition of $\hat{A}$, using the fact that the eigenvectors of W are the same as those of L_{sym} and the relationship between the eigenvalues of W and L_{sym} are $\Lambda = 1 - \Lambda_W$, where Λ_W and Λ are those of W and L_{sym}, respectively. Figure 1 explains in detail the procedure to approximate L_{sym} using Nyström method.

Proposed Nyström approximation of the normalized Laplacian L_{sym}

Input : adjacency matrix $W \in \mathbf{R}^{N \times N}$, the number of columns to be sampled c

Output : The approximate eigenvalues $\tilde{\Lambda}$ and eigenvectors $\tilde{U} \in \mathbf{R}^{N \times c}$ of L_{sym}

1. $C \leftarrow W(\ :\ , \Omega)$

2. $A \leftarrow C(\ \Omega, :\ )$, $B \leftarrow C(\Omega^c, :\ )$

3. $\hat{A} \leftarrow D_A^{-1/2} A D_A^{-1/2}$, $D_A(i,i) = \sum_j A(i,j)$

4. $\hat{B} \leftarrow D_B^{-1/2} B \Delta_B^{-1/2}$, where $D_B(i,i) = \sum_j A(i,j)$, $\Delta_B(i,i) = \frac{c}{N-c} \sum_i A(i,j)$

5. spectral decomposition : $\hat{A} = U_{\hat{A}} \Lambda_{\hat{A}} U_{\hat{A}}^T$

6. $\hat{C} \leftarrow \begin{bmatrix} \hat{A} \\ \hat{B} \end{bmatrix}_{\Omega}$, rearrange in the reverse order Ω ..

7. $\tilde{\Lambda} = 1 - \frac{N}{c} \Lambda_{\hat{A}}$, $\tilde{U} = \sqrt{\frac{c}{N}} \hat{C} U_{\hat{A}} \tilde{\Lambda}^{\dagger}$

Figure 1. The proposed Nyström approximation of the normalized Laplacian.

4. Experiments

Patch sets associated with signals are constructed. Commute time embedding is performed on the patch set composed of N patches, each vertex is mapped into three dimensional vector, as shown in Equation (5) to get a feel for what the dataset looks like [7].

4.1. Investigating the characteristics of commute time embedding

In figure 2, we show commute time embedding of some sinusoid. The segment is composed of 700 samples, from which 676 patches are extracted so that they may be maximally overlapped. In this figure, patches of lower variance are encoded with blue colour, while patches of higher variance with red colour.

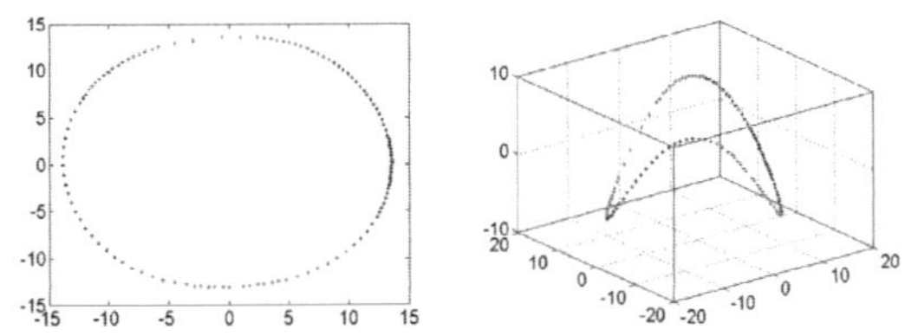

Figure 2. Commute time embedding of a sinusoidal signal from which the patch graph is constructed. Left: Two-dimensional Embedding, Right: Three-dimensional embedding.

As mentioned in the previous section, commute time embedding can be used as a tool for geometrization of dataset, where some periodic phenomena are transformed into the geometries composed of topological circles or higher dimensional holes in some space.

4.2. The proposed approximate commute time embedding using Nyström method

Figure 3 shows the commute time embedding results of the patch graph same as that of figure 2, using Nyström methods. As mentioned in the above section, severe distortion occurs on the embedding results of Choromanska's method, compared with those of our proposed method. Irregular scales at the embeddings, which is caused by the errors in approximating the eigenvalues of L_{sym}, seems to degrade the performance. Note that the phase can be reversed because the embedding map consists of eigenvectors, as given in equation (5). Table 1 shows the five smallest eigenvalues of the L_{sym} for the patch graph of sinusoid. According to equation (6), the smallest eigenvalue of L_{sym} for the connected graph should be zero and the largest one be less than 2. Choromanska's method, as given in table 2, does not satisfy the conditions. However, table 3 shows that the approximate eigenvalues computed using our proposed method approach true values as the number of samples increases as well as satisfy the condition given in equation (6).

Table 1. The five smallest eigenvalues of L_{sym}.

	λ_1'	λ_2'	λ_3'	λ_4'	λ_5'
values	0	0.01	0.012	0.040	0.045

Table 2. The five smallest eigenvalues of L_{sym}, obtained using the Choromanska's method.

#samples	λ_1'	λ_2'	λ_3'	λ_4'	λ_5'
200	2.304	2.331	2.340	2.359	2.425
400	0.637	0.650	0.711	0.729	0.745
600	0.119	0.131	0.136	0.166	0.175

IOP Conf. Series: Journal of Physics: Conf. Series **814** (2017) 012010 doi:10.1088/1742-6596/814/1/012010

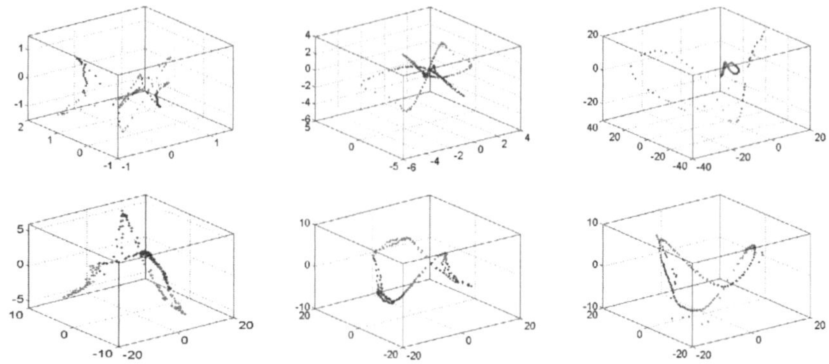

Figure 3. Approximate commute time embedding using Nyström method. The upper row: Choromanska's method, the lower row: our proposed method, Sampling rates: The first column (29.6%), the second column (59.2%), the third column (88.7%).

Table 3. The five smallest eigenvalues of L_{sym}, obtained using our proposed method.

#samples	λ_1'	λ_2'	λ_3'	λ_4'	λ_5'
200	0.0	0.006	0.009	0.027	0.036
400	0.0	0.009	0.011	0.038	0.041
600	0.0	0.010	0.011	0.040	0.046

5. Conclusions

This paper has explored approximation of commute time embedding to reduce the computational burden for large dataset. Our proposed method is based on Nyström sampling method to compute approximate commute time. The strength of our method is that it preserves the properties that the normalized graph Laplacian matrix is SPSD even though it is approximated via sampling process. Thus, our method can be applied efficiently to dimensionality reduction, which is very effective for visualization, as well as spectral clustering or pattern classification of large dataset.

As a future research, we would like to explore its application to pattern classification or manifold visualization in a geometric way.

6. References

[1] Tenenbaum J, deSilva V and Langford, J 2000 *Science vol 290* pp 2319-2323
[2] Belkin M and Niyogi P 2003 *Neural Computation vol 15 no 6* pp 1373-1396
[3] Qiu H and Hancock E 2007 *IEEE Trans. PAMI, vol 29 no 11* pp 1873-1890
[4] Williams C and Seeger M 2001 *In Annual Advances in Neural Information Processing Systems 13: Proceeding of the 2000 Conference* pp 682-688
[5] Drineas P and Mahoney M 2005 *Journal of Machine Learning Research 6*, pp 2153-2175
[6] Choromanska A, Jebara T, Kim H, Mohan M and Monteleoni C 2013 *ALT2013, LNAI 8139* pp 367-381
[7] Taylor K 2011 *PhD Thesis, University of Colorado, Boulder, Dept. of Applied Mathematics*

Comparison of combustion products by the iteration method and application of gasoline and biogas

Paramust Juntarakod

Department of Mechatronic Engineering, Faculty of Engineering
Rajamangala University of Technology Isan, Khonkaen Campus.150 Srichan Road, NaiMuang, Muang, Khonkaen, 40000, Thailand

paramust_kmitnb@hotmail.com

Abstract. For a specific combustion problem involving calculations of several species at the equilibrium state, it is simpler to write a general computer program and calculate the combustion concentration. Original work describes, an adaptation of Newton-Raphson method was used for solving the highly linear system of equations describing the formation of equilibrium products in reacting of fuel-additive-air mixtures. This study also shows what possible of the results. In this paper, it presents the efficient numerical algorithms for solving the combustion problem, to be used linear equations based on the iteration method and high order of the Taylor series. The modified the Adomian decomposition method was applied to construct the numerical algorithms. Some numerical illustrations are given to show the efficiency of algorithms. Comparisons of results by the new Matlab routines and previous routines, the result data indicate that the new Matlab routines are reliable and application with gasoline and biogas with the variance of equivalence ratios, typical deviations from previous results are less than 0.1%.

1. Introduction

Chemical equilibrium is a condition when the chemical reaction, by which individual species in the burned gases react together, produces and removes each species at equal rates; hence at this condition has no net change in species composition results. Chemical equilibrium is usually described by either of two equivalent formulations, namely the equilibrium constants method (ECM) and the minimization of free energy (Δg =0). However, the two formulations reduce to the same number of iteration equations.

When considering a general chemical equilibrium, there are several disadvantages in the equilibrium constant method such as the system of chemical equilibrium which are linear equations, more data and recoding. Including are numerical difficulties with the use of components and more difficulties in solving. On the other hand, the equilibrium constant method is simpler to adopt and formulate for a simple situation, and it was found that for most combustion problems, it gives satisfactory results without elaborate and complicated computer programs.

Mohamed[1] and **Caton** [2] developed procedures that combines algebraically the equilibrium equations, eliminates some unknowns, and scales the simplified equations to obtain a set of simultaneous linear equations that are solved with linear algebra. Their procedures combine the Gauss-Seidel and Newton methods. They claim that the Gauss-Seidel is unreliable and the Newton-Raphson method is slow.

When considering a minimization of free energy, the application of approaches results in a system of non-linear equations which can be solved by different numerical methods **Adnan**[3]. The system uses the Newton-Raphson method for solving. The initial values of the independent variables and the Lagrangian multipliers used in the calculations are determined on the basis of a numerous calculations of the composition of gaseous, liquid, and solid fuel combustion products.
Buttworth[4], an effort is made by developing a combustion mathematical model to simulate exhaust emissions in fuels. Based on Equilibrium Constants Method (ECM), a computer program using MATLAB has been developed for the fuels to calculate the mole fractions of the emission gases. ECM is based on thermodynamic measurements and empirical calculations. Thermodynamic data for elements, combustion products and many pollutants are available in a compilation published by the National Bureau of Standards, called the JANAF (Joint Army–Navy–Air Force) tables. The equilibrium constant data from this tables using polynomial curve fitted has been used in calculating the combustion products.
Ferguson[5], a solution for the properties of equilibrium combustion products based on an equilibrium constant method (ECM) with the gas phase products of combustion of hydrocarbon fuels. The use of equilibrium constants is based on minimization of the Gibb free energy of gas. The equilibrium constant method is simpler than the Lagrange multiplier approach when considering restricted species lists. However, the equilibrium constant method requires the equilibrium reactions, such the water-gas reaction.
Solving system of linear equations of combustion is one of the most important problems in numerical analysis. Much attention has given to develop several iterative methods for solving linear equations. **Babolian[6]** have proposed and studied several methods for linear equations with higher order convergence by using the decomposition technique. He has to use the higher order differential derivatives which are a serious drawback. To overcome this drawback, we suggest and analyze a method for solving linear equations, which does not been involved the higher order derivative of the function.
Scientists and engineers have adapted the application of iterative methods in linear algebraic equations such as Newton–Raphson method, Adomian decomposition method and Householder iteration method lately. It is because these methods simplify complex problems into simple solutions. Newton–Raphson method is the most popular and powerful method in solving non-linear combustion equations, for example, **Rakopoulos**[7] also used this method to solve equations of chemical equilibrium combustion model in 11 species diesel alone combustion products, **Masood**[8] used this method to solve equations of low-temperature combustion model with only 10 products species , and **Rashidi** [9] combined Newton–Raphson with successive substitution methods to solve combustion problem of hydrocarbon fuel with 18 products species and concluded that the combined method gives fast and reliable convergence during iteration process.
In the present study, it suggests and analyzes iterative methods for solving linear equations. And it used an effort made by developing a combustion mathematical model to simulate exhaust emissions in hydrocarbon fuel as Biogas and Gasoline. Based on Equilibrium Constants Method, a computer program using MATLAB has been developed for the fuels to calculate the mole fractions of the emission gases. ECM is based on thermodynamic measurements and empirical calculations. Thermodynamic data for elements, combustion products and many pollutants are available in a compilation published by **Grill** [10].

2. Methods

2.1 Problem formation

This paper involves the simulation for determining and comparing Newton–Raphson method [NRM], Householder's iteration method [HoIM], Adomian Decomposition method [ADM] and High order iteration Method [HIM] for the mole fraction of each of the exhaust species when the hydrocarbon fuel is burnt along with air and the results are presented. The proportion of hydrocarbon in the hydrocarbon–air blend affecting the mole fraction of the exhaust species is also simulated. Numerical experiments investigations were carried out, in gasoline fuel mode, which showed a good agreement between the predicted and experimental results. Here in this paper an effort is made to calculate the exhaust emissions in gasoline fuel mode. The optimal combination of fuels depending upon the exhaust can be found effectively and applied practically.

The fuel is to be specified in terms of the C, H, O and N atoms in the fuel. For the blend of fuel-air considered. The other parameters that need to be specified are equivalence ratio, pressure and temperature. For the calculation of equilibrium constant, the data for constants is considered from **Grill** [10] tables. The molar-air fuel ratio is calculated from the number of carbon, hydrogen, nitrogen and oxygen atoms present in the fuel.

2.2 Logic of the numerical simulation

The basis of the equilibrium combustion products with fuel-air mixture model is a solution to the atom balance equations from the chemical reaction equation of fuel and air forming. This mixture equation is given in Eq-(1) for the condition of equivalence ratio ($\phi < 3$). This process based on conservation of mass that was modified form **Ferguson**[5], where the following equation represents the chemical reaction with the relevant species involved which was added gasoline supplied as given below,

$$\varepsilon\phi\left(\chi_1 C_{\alpha 1} H_{\beta 1} O_{\gamma 1} N_{\delta 1} + \chi_2 C_{\alpha 2} H_{\beta 2} O_{\gamma 2} N_{\delta 2}\right) + 0.21 O_2 + 0.79 N_2 \rightarrow y_1 CO_2 + y_2 H_2O + y_3 N_2 + y_4 O_2 + y_5 CO + y_6 H_2 + y_7 H + y_8 O + y_9 OH + y_{10} NO + y_{11} N \quad (1)$$

Where $\varepsilon = \dfrac{0.210}{(\alpha 1 + \alpha 2) + 0.25(\beta 1 + \beta 2) - 0.5(\gamma 1 + \gamma 2)}$

There is the conservation of 4 atoms and from mixture equation, so atom balancing can be written as

$$C \quad \varepsilon\phi(\alpha 1 + \alpha 2) = y_1 + y_2 \quad (2)$$

$$H \quad \varepsilon\phi(\beta 1 + \beta 2) = 2y_1 + 2y_6 + y_7 + y_9 \quad (3)$$

$$O \quad \varepsilon\phi(\gamma 1 + \gamma 2) + 0.42 = 2y_1 + y_2 + 2y_4 + y_5 + y_8 + y_9 + y_{10} \quad (4)$$

$$N \quad \varepsilon\phi(\delta 1 + \delta 2) + 1.5 = 2y_3 + y_{10} + y_{11} \quad (5)$$

Also the constraint that the mole fractions add up to unity must be satisfied,

$$y_1 + y_2 + y_3 + y_4 + y_5 + y_6 + y_7 + y_8 + y_9 + y_{10} + y_{11} - 1 = 0 \quad (6)$$

The expression for atom balance of each equation can be eliminated by dividing Eq-(2) by Eq-(3). The equation can be written as next equation.

$$2y_2 + 2y_6 + y_7 + y_9 - d_1(y_1 + y_5) = 0 \quad (7)$$

Likewise the Eq-(3) and Eq-(4), the equation can be written as

$$2y_1 + y_2 + 2y_4 + y_5 + y_8 + y_9 + y_{10} - d_2(y_1 + y_5) = 0 \quad (8)$$

Likewise the Eq-(1) and Eq-(4), the equation can be written as

$$2y_2 + y_{10} + y_{11} - d_3(y_1 + y_5) = 0 \quad (9)$$

Equations of constant value for a simple studied can be defined as,

$$d_1 = \frac{\beta 1 + \beta 2}{\alpha 1 + \alpha 2}, d_2 = \frac{\gamma 1 + \gamma 2}{\alpha 1 + \alpha 2} + \frac{0.42}{\varepsilon\phi(\alpha 1 + \alpha 2)}, d_3 = \frac{\delta 1 + \delta 2}{\alpha 1 + \alpha 2} + \frac{1.58}{\varepsilon\phi(\alpha 1 + \alpha 2)}$$

The above 4 equations (Eq-(6) through Eq-(9)) have 11 unknowns ($y_1 \rightarrow y_{11}$), therefore in order to solve for these 11 unknowns other 7 more equations are needed which may be derived from the consideration of equilibrium among products. The equilibrium constant can be related to the partial pressure of the reactants and products as show Table 1.

Table 1. The partial pressure of the reactants and products and the equilibrium constant equations

$CO_2 \leftrightarrow CO + (1/2)O_2$	$K_1 = y_5 y_4^{1/2} P^{1/2} / y_1$
$H_2 + (1/2)O_2 \leftrightarrow H_2O$	$K_2 = y_2 / y_4^{1/2} y_6^{1/2} P^{1/2}$
$(1/2)H_2 + (1/2)O_2 \leftrightarrow OH$	$K_3 = y_9 / y_4^{1/2} y_6^{1/2}$
$(1/2)H_2 \leftrightarrow H$	$K_4 = y_7 P^{1/2} / y_6^{1/2}$
$(1/2)O_2 \leftrightarrow O$	$K_5 = y_8 P^{1/2} / y_4^{1/2}$
$(1/2)N_2 \leftrightarrow N$	$K_6 = y_{11} P^{1/2} / y_3^{1/2}$
$(1/2)O_2 + (1/2)N_2 \leftrightarrow NO$	$K_7 = y_{10} / y_4^{1/2} y_3^{1/2}$

And the partial pressure of a component is defined relative to the total pressure and the mole faction, thus the equilibrium constant can be rewritten as,

Table 2. The equilibrium constant of constant term

$y_1 = C_1 y_5 y_4^{1/2}$	$C_1 = P^{1/2} / K_1$
$y_2 = C_2 y_4^{1/2} y_6^{1/2}$	$C_2 = K_2 P^{1/2}$
$y_9 = C_3 y_4^{1/2} y_6^{1/2}$	$C_3 = K_3$
$y_7 = C_4 y_6^{1/2}$	$C_4 = K_4 / P^{1/2}$
$y_8 = C_5 y_4^{1/2}$	$C_5 = K_5 / P^{1/2}$
$y_{11} = C_6 y_3^{1/2}$	$C_6 = K_6 / P^{1/2}$
$y_{10} = C_7 y_4^{1/2} y_3^{1/2}$	$C_7 = K_7$

The equilibrium constants are treated as a function of gas temperature T (K). It is the highest possible temperature that can be achieved during combustion. For computation of temperature, it is assumed that no heat will be transferred through combustion chamber walls, that is heat must be zero and all energy transferred to engine work and exhaust product. Values for K have been tabulated for several reactions at various temperatures. These values come from the above relations, which are determined by statistically thermodynamics. Equilibrium constant in Table 2 are curve fitted to the Thermo chemical **Grill[10]** and their expressions are,

$$K_i = \exp\left((a_1)(\ln T) + \frac{a_2 T}{2} + \frac{a_3 T^2}{6} + \frac{a_4 T^3}{12} + \frac{a_5 T^4}{20} + \frac{a_6}{2} + a \right) \tag{10}$$

Where T is temperature (Kelvin). For the range of $3500 \le T \le 5000K$, coefficients for calculating the equilibrium constant K_i, Substitution of the equilibrium constants from Eq-(10) and the curve-fit coefficient constants into Table 1 and rearranging to express mole factions of all species in terms of y_3,y_4,y_5 and y_6, the mole factions of CO,H_2,O_2 and N_2 respectively. The expressions resulted are four linear equations in four unknowns are as follows,

$$f_i(y_3, y_4, y_5, y_6) = 0 \tag{11}$$

Where,

$$C_1 y_5 y_4^{1/2} + C_2 y_4^{1/2} y_6^{1/2} + y_3 + y_4 + y_5 + y_6 + C_4 y_6^{1/2} + C_5 y_4^{1/2} + C_3 y_4^{1/2} y_6^{1/2} + C_7 y_4^{1/2} y_3^{1/2} + C_6 y_3^{1/2} - 1 = 0 \quad (12)$$

$$2C_2 y_4^{1/2} y_6^{1/2} + 2y_6 + C_4 y_6^{1/2} + C_3 y_4^{1/2} y_6^{1/2} - d_1\left(C_1 y_5 y_4^{1/2} + y_5\right) = 0 \quad (13)$$

$$2C_1 y_5 y_4^{1/2} + C_2 y_4^{1/2} y_6^{1/2} + 2y_4 + y_5 + C_5 y_4^{1/2} + C_3 y_4^{1/2} y_6^{1/2} + C_7 y_4^{1/2} y_3^{1/2} - d_2\left(C_1 y_5 y_4^{1/2} + y_5\right) = 0 \quad (14)$$

$$2C_2 y_4^{1/2} y_6^{1/2} + C_7 y_4^{1/2} y_3^{1/2} + C_6 y_3^{1/2} - d_3\left(C_1 y_5 y_4^{1/2} + y_5\right) = 0 \quad (15)$$

2.3 Numerical solution using modified iteration method for Combustion equation
In this study, Newton–Raphson method, Householder's iteration method, Adomian Decomposition method and High order iteration Method were selected to solve equilibrium constants combustion equations of 11 combustion products of gasoline-air blend, which were summarized in Eqs. (12)–(15). This system can be written in more compact vector from as

$$F(y) = \begin{bmatrix} f_1(y_3, y_4, y_5, y_6) \\ f_2(y_3, y_4, y_5, y_6) \\ f_1(y_3, y_4, y_5, y_6) \\ f_1(y_3, y_4, y_5, y_6) \end{bmatrix} \quad (16)$$

The iteration method requires the evaluation of a matrix, known as the Jacobian (J) of the system. The that system is the elements of the Jacobian matrix, J for the previously derived combustion reaction equations are shown in Eq-(12) until Eq.-(15). The stated values were obtained from mole fractions of these species in atmospheric air at room temperature.
The solution was unique as long as mole fractions of all 11 products species showed positive values. By using the stated initial guess, most of the results indicated almost constant values after tenth iterations, which shows that there were no convergence problems. Multiple solutions will be detected if the result showed negative value of mole fraction for one or more products that was due to wrong selection of initial guess. Furthermore, there was no result shown after 1000 iterations if divergence occurred.

2.3.1 Newton–Raphson method. The given equations can be solved by using the Newton Raphson method. The general form of Newton Raphson method for linear systems is

$$Y_{K+1} = Y_K - \frac{f(x_k)}{f'(x_k)} \quad \text{k=Iteration} \quad (18)$$

Each of these may be expanded in Taylor expansion of $f(x)$ to a second order for Newton Raphson method,

$$f_j + \frac{\partial f_j}{\partial y_3}\Delta y_3 + \frac{\partial f_j}{\partial y_4}\Delta y_4 + \frac{\partial f_j}{\partial y_5}\Delta y_5 + \frac{\partial f_j}{\partial y_6}\Delta y_6 \approx 0,\ j = 1,2,3,4$$

(19)

Functions f_j are evaluated from the solution of interested functions (Eq-(12) through Eq-(15). $f_j = f_{(1,2,3,4)}$ The independent set of derivatives is obtained by solution of matrix equation that results

IOP Conf. Series: Journal of Physics: Conf. Series **814** (2017) 012011 doi:10.1088/1742-6596/814/1/012011

from differentiating with respect to mole faction. The above can be arranged as set of linear equations in the matrix form, $[A][Z]-[B]=0$, where $[A]=\left[\frac{\partial f}{\partial y}\right],[B]=[f_{ij}],[Z]=[\Delta y]$

2.3.2 Householder's iteration method. The given equations can be solved by using Householder's iteration method. The general form of Householder's iteration method for linear systems

$$\left(f(x_k)+f^2(x_k)f^{''}(x_k)f^{'-2}(x_k)\right)+f^{'}(x_k)\left(Y_{K+1}-Y_K\right)=0$$
(20)

Each of these may be expanded in Taylor expansion of $f(x)$ to a higher order for Householder's iteration method. That is,

$$f_j=f(x_k)+f^2(x_k)f^{''}(x_k)f^{'-2}(x_k)$$

2.3.3 The Adomian Decomposition method. The given equations can be solved by using Adomian Decomposition method. The general form of Adomian Decomposition method for linear systems is

$$f(x_k)+f^2(x_k)f^{''}(x_k)f^{'-2}(x_k)+\frac{f^3(x_k)f^{''2}(x_k)f^{'-4}(x_k)}{2}+f^{'}(x_k)\left(Y_{K+1}-Y_K\right)=0 \quad (21)$$

Each of these may be expanded in Taylor expansion of $f(x)$ to a higher order for Adomian Decomposition method, that is

$$f_j=f(x_k)+f^2(x_k)f^{''}(x_k)f^{'-2}(x_k)+\frac{f^3(x_k)f^{''2}(x_k)f^{'-4}(x_k)}{2} \quad (22)$$

2.3.4 High order iteration method. The given equations can be solved by using High order iteration method. The general form of High order iteration method for linear systems is

$$f(x_k)+f^2(x_k)f^{''}(x_k)f^{'-2}(x_k)+\frac{f^3(x_k)f^{'''}(x_k)f^{'-3}(x_k)}{6}+f^{'}(x_k)\left(Y_{K+1}-Y_K\right)=0$$
(23)

Each of these may be expanded in Taylor expansion of $f(x)$ to a higher order for High order iteration method.

$$f_j=f(x_k)+f^2(x_k)f^{''}(x_k)f^{'-2}(x_k)+\frac{f^3(x_k)f^{'''}(x_k)f^{'-3}(x_k)}{6}$$
(24)

2.4 Matlab inputs

In this study, Matlab program code for the combustion simulation has been developed. The input parameters such as Biogas+Air ($C_5H_7O_2N+0.21O_2+0.79N_2$), Gasoline+Air ($C_7H_{17}+0.21O_2+0.79N_2$) and Gasoline 50%+Biogas 50% by mole with air ($0.5C_7H_{17}+0.5\ C_5H_7O_2N+0.21O_2+0.79N_2$), respectively, after that substitution in term of equivalence ratio such as 0.5 to 2.0. Assuming the invariant adiabatic temperatures were 2,000-3,000 K with constant pressure was 50 atm and the difference of convergence was $1.0\text{x}10^{-10}$.

3. Results and discussion

This section presents summarized Eq-(12) through Eq-15 to illustrate the efficiency of the iterative method proposed in this paper. We compare the method [denoted by Iteration method] with Newton-

Raphson method (NRM), Householder's iteration method (HoIM), Adomian Decomposition method (ADM), and High order iteration Method (HIM). The results were compared the number of iteration in each method obtained solution or mole faction of combustions and difference value from Newton-Raphson method (NRM-HoIM, NRM-ADM and NRM-HIM).

Numerical computations have been carried out by using the software Matlab 2010. The results are presented in Tables 3 for comparing with Newton-Raphson method of the mole fractions of H_2O, CO_2 and equivalence ratio 0.8-1.2, temperature 2000 K and 3000 K.

At low adiabatic temperature, 2000 K, the number of iterations of High order are less than the old method of Newton Raphson method in Table below, but the solutions computed by the new method are more exact and similar values. At high adiabatic temperature, 3000 K, the number of iterations of High order are more than the old method of Newton Raphson method in Table below, but the solutions computed by the new method are more exact and similar values.

Table 3. The obtained solution of the mole factions of H_2O, CO_2

	equivalence ratio 0.8		equivalence ratio 1.0		equivalence ratio 1.2	
Method	obtained solution	Iteration	obtained solution	Iteration	obtained solution	Iteration
Mole fraction of H_2O, 2000 K						
NRM	0.088661111630029	145	0.106774699970414	146	0.123609519024991	147
HoIM	0.088661134281620	145	0.106774724763846	146	0.123609533587120	146
ADM	0.088661134471458	145	0.106774724996680	146	0.123609533720466	146
HIM	0.088661134473316	145	0.123609533721766	146	0.123609533721766	145
NRM-HoIM	-0.000000022651591		-0.000000024793432		-0.000000014562129	
NRM-ADM	-0.000000022841429		-0.000000025026266		-0.000000014695475	
NRM-HIM	-0.000000022843286		-0.016834833751352		-0.000000014696775	
Mole fraction of CO_2, 2000 K						
NRM	0.078560867873312	145	0.099545606322872	146	0.119867071850334	147
HoIM	0.078560894228601	145	0.099545631832484	146	0.119867095948412	146
ADM	0.078560894478903	145	0.099545632086899	146	0.119867096194587	146
HIM	0.078560894481443	145	0.119867096197060	146	0.119867096197060	145
NRM-HoIM	-0.000000026355289		-0.000000025509612		-0.000000024098077	
NRM-ADM	-0.000000026605591		-0.000000025764027		-0.000000024344252	
NRM-HIM	-0.000000026608131		-0.020321489874188		-0.000000024346726	
Mole fraction of H_2O, 3000 K						
NRM	0.088690819062502	144	0.107316922189171	146	0.124697861172053	141
HoIM	0.088690844446441	188	0.107316947586183	190	0.124697887443739	185
ADM	0.088690844692973	231	0.107316947835203	233	0.124697887692734	229
HIM	0.088690844695659	297	0.124697887694876	295	0.124697887694876	295
NRM-HoIM	-0.000000025383938		-0.000000025397012		-0.000000026271687	
NRM-ADM	-0.000000025630471		-0.000000025646032		-0.000000026520681	
NRM-HIM	-0.000000025633157		-0.017380965505705		-0.000000026522823	
Mole fraction of CO_2, 3000 K						
NRM	0.039869692159035	144	0.054955151813055	146	0.071742451241782	141
HoIM	0.039869706649732	188	0.054955163498001	190	0.071742472785569	185
ADM	0.039869706806835	231	0.054955163622364	233	0.071742473008710	229
HIM	0.039869706808605	297	0.071742473010849	295	0.071742473010849	295
NRM-HoIM	-0.000000014490697		-0.000000025397012		-0.000000021543787	
NRM-ADM	-0.000000014647799		-0.000000025646032		-0.000000021766928	
NRM-HIM	-0.000000014649570		-0.017380965505705		-0.000000021769067	

The results illustrating composition with temperature and equivalence ratio are given in figure 1 to 3 for the combustion of $C_5H_7O_2N$ (Biogas), C_7H_{17} (Gasoline) and gasoline-biogas blend as 50% by mole at P=50 bar. The largest mole fractions are N_2, H_2O and CO_2. At this pressure, the composition predicted by using Newton Raphson method is a good approximation for all temperature 2000 K . At low pressure, dissociation is even greater. So that, at atmospheric pressure, Newton-Raphson method is low order iteration method which are valid for all temperature. The clean illustrating results was an experiential rise in product species such as CO, NO, OH, O_2, O, H_2 and H. For lean $\phi < 1$ condition, the O_2 fraction is relatively insensitive to temperature. For rich condition, the H_2 mole fraction first decreases, then increases with increasing temperature.

4. Conclusions

The accuracy of the MATLAB program equilibrium combustion products routine was established through comparison with a comparable MATLAB program routine (**Buttsworth**[5]). Maximum differences in molar concentrations of less than 0.01% were typically apparent over the range of conditions calculated by the four routines (NRM, HoIM, ADM and HIM). Adiabatic fame temperature calculations (1500-3000 K) also confirmed that differences between the new Matlab routines and established results are typically less than the round-off errors in the established results around 0.1%.

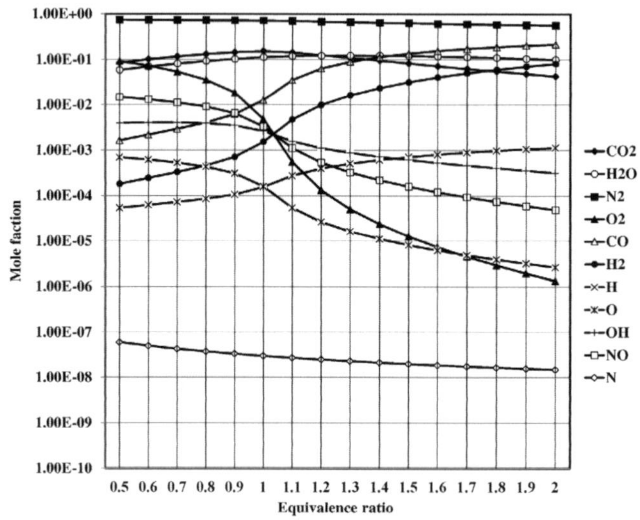

Figure 1. Mole fraction of Biogas combustion products at Temperature 2000 K, Pressure 50 Bars

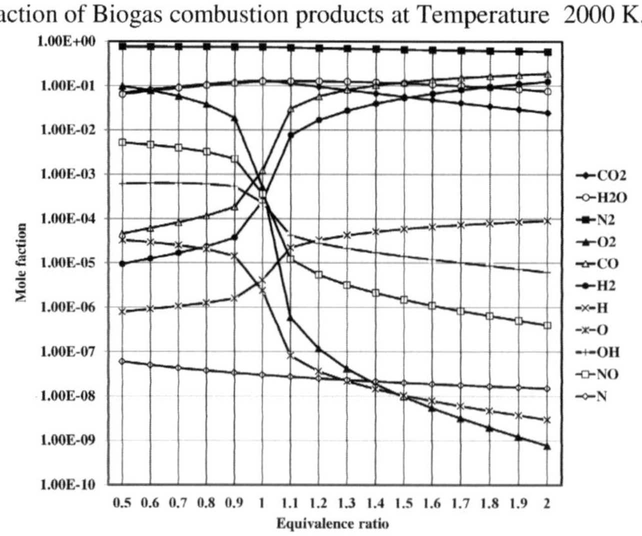

Figure 2. Mole fraction of Gasoline combustion products at Temperature 2000 K, Pressure 50 Bars

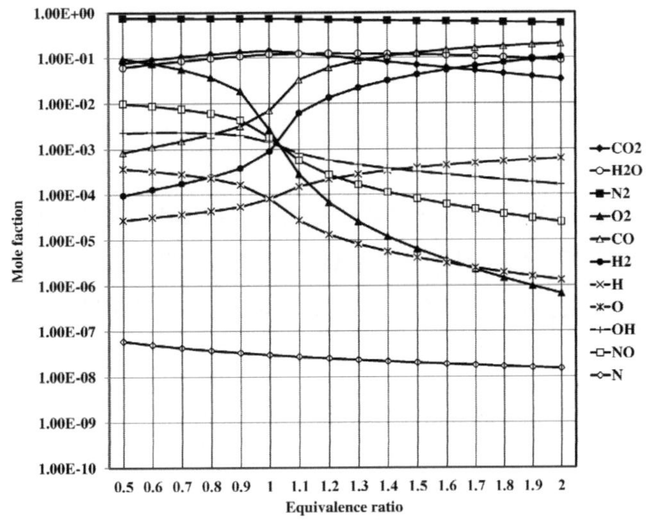

Figure 3. Mole fraction of Gasoline (50%) and Biogas (50%) combustion products at Temperature 2000 K, Pressure 50 Bars

5. References

[1] M. Mohamed, E. Awad, Determining the Thermodynamic Properties of Fuel Blends for Use with Air-Fuel Models of IC Engines. in: International Conference on Trends in Industrial and Mechanical Engineering, Dubai, UAE, (ICTIME'2012) March 24-25, 2012.

[2] J.A. Caton, The thermodynamic characteristics of high efficiency, Internal Combustion Engines, Energy Conversion and Management 58 (2012).3.

[3] R. Adnan. H.H. Masjuki. T.M.I. Mahlia. Mathematical modelling on the effect of equivalence ratio in emission characteristics of compression ignition engine with hydrogen substitution, Applied Mathematics and Computation 217 (2011), 6144–6158.

[4] D. R. Buttsworth, Spark Ignition Internal Combustion Engine Modelling using Matlab, Faculty of Engineering & Surveying Technical Reports, University of Southern Queensland [online]http://eprints.usq.edu.au/173/1/tr-2002-02.pdf. (accessed Nov 1, 2011)

[5] C. R. Ferguson, Internal Combustion Engines, Applied Thermo sciences, John Wiley and Sons, New York, USA,1986.

[6] E. Babolian, J. Biazar, Solution of linear equations by modified Adomian decomposition method, Applied Mathematics and Computation 132 (2002), 167–172.

[7] C. D. Rakopoulos, D. T. Hountalas, E. I. Tzanos, G. N. Taklis, A fast algorithm for calculating the composition of diesel combustion products using 11 species chemical equilibrium scheme, Advances in Engineering Software 19 (1994) 109–119.

[8] M. Masood, M.M. Ishrat, Computer simulation of hydrogen–diesel dual fuel exhaust gas emissions with experimental verification, Fuel 87 (2008) 1372–1378.

[9] M. Rashidi, Calculation of equilibrium composition in combustion products, Chem. Eng. Technol 20 (1997) 571–575.

[10] M. Grill, M. Chiodi. H. Berner, M. Bargende. Calculating the Thermodynamic Properties of Burnt Gas and Vapor Fuel for User-Defined Fuels, MTZ 05 /2007 Volume 68 [online] http://www.all4engineers.de/animationen/m05-07-14.pdf (accessed Nov 1, 2011)

Acknowledgments
Faculty of Engineering. Rajamangala University of Technology Isan, Khonkaen Campus.150 Srichan Road, NaiMuang, Muang, Khonkaen, 40000, Thailand.

IOP Conf. Series: Journal of Physics: Conf. Series **814** (2017) 012012 doi:10.1088/1742-6596/814/1/012012

Integrated optimisation technique based on computer-aided capacity and safety evaluation for managing downstream lane-drop merging area of signalised junctions

CHAI Chen[1], and WONG Yiik Diew[1]

[1]Center for Infrastructure System, Nanyang Technological University, Singapore

chai0076@e.ntu.edu.sg

Abstract. This study provides an integrated strategy, encompassing microscopic simulation, safety assessment, and multi-attribute decision-making, to optimize traffic performance at downstream merging area of signalized intersections. A Fuzzy Cellular Automata (FCA) model is developed to replicate microscopic movement and merging behavior. Based on simulation experiment, the proposed FCA approach is able to provide capacity and safety evaluation of different traffic scenarios. The results are then evaluated through data envelopment analysis (DEA) and analytic hierarchy process (AHP). Optimized geometric layout and control strategies are then suggested for various traffic conditions. An optimal lane-drop distance that is dependent on traffic volume and speed limit can thus be established at the downstream merging area.

1. Introduction

Whenever several road lanes are merged to continue into fewer lanes, a merging area is defined. These merging areas, at expressway on-ramp as well as at lane-drop (decrease in number of lanes) zone at signalized junctions, are critical bottle-necks due to lesser capacity by virtue of fewer numbers of the merged lanes [1]. Moreover, as vehicles jostle with each other along the merging area in moving into the fewer lane(s), the safety aspect at the merging area is also an important consideration.

Vehicle merging man oeuvres can be classified as: 1) priority merging; and 2) non-priority merging [2]. Priority merging typically occurs when higher-priority major stream vehicles merge with lower-priority stream vehicles. Such merging rules are also defined as give-way on-ramp merging [3]. On the other hand, non-priority merging includes lane-drop or on-ramp merging without priority between competing traffic streams. For priority merging, there is one bottle-neck zone (at approach area of the minor stream) whereas for non-priority merging there are two bottle-neck zones (at the approaches of each stream). The dynamics of vehicle movements at merging areas have been modelled by several studies at microscopic level [4,5] and at macroscopic level [6]. It is found from these simulation models that capacity as well as safety at the merging areas will decrease when there is a higher vehicle density, shorter merging area and higher operating speed [2,7].

Existing modelling techniques of the merging movements have not been able to provide comprehensive analysis of driving behaviour and vehicle interactions at the merging areas. In essence, most microscopic and macroscopic simulations are essentially confined to the physical movements. On the other hand, behavioural aspects such as cognitions and strategic decisions of individual drivers have seldom been successfully modelled. Certainly, traffic conflicts from competing traffic streams at the merging area result in greater accident risk, and improving traffic safety is always a critical concern [8]. However, most safety studies are based on incidences of crashes. Herein, an integrated approach for both capacity and safety assessment at merging area, especially at the planning stage, is well warranted.

This study provides an integrated strategy encompassing microscopic simulation, safety assessment, and multi-attribute decision-making to optimize traffic performance at the downstream merging area of signalized junctions. A Fuzzy Cellular Automata (FCA) model is developed to replicate microscopic movement and merging behaviour at the merging area. Notably, fuzzy logic is capable of handling uncertainties which are well suited for representing multiple co-operations, collaborations and even conflicts [9, 10]. Moreover, compared to other simulation models, linguistic terms are used to describe environment and responses in the FCA model. These linguistic terms can be used to describe drivers' cognitions, such as perception, intention and attitude. In this way, decision-making process of individual vehicles can be modelled in a very clear and straight-forward way. Furthermore, as behaviour of an individual can be controlled by more than one fuzzy set, complex behaviour of merging traffic can be modelled. Compared to traditional Cellular Automata (CA) models, the proposed FCA model is incorporated with fuzzy logic and accounts for driving cognitions and decision-making of individual drivers [11].

Simulation experiments based on FCA are conducted in different scenarios of various traffic conditions and ramp layout at typical geometric layouts of the merging area. Simulation outputs include characteristics on both capacity and safety aspects, such as travel time, traffic delay, average velocity, occurrences and severity of conflicts. The results are then evaluated through the data envelopment analysis (DEA) and analytic hierarchy process (AHP). Optimized geometric layout and control strategies are then suggested for various traffic conditions [12].

2. Fuzzy Cellular Automata (FCA) model for merging traffic

Please follow these instructions as carefully as possible so all articles within a conference have the same style to the title page. This paragraph follows a section title so it should not be indented.

2.1. Structure and elements of merging traffic

A system of merging traffic, including both on-ramp and lane-drop merging, has three major components: namely arriving flow, merging area and exiting flow. The definition of merging area in this study is the location where vehicles change lane due to ensuing change of lane configuration by way of fewer lane counts at or of some distance before the onset of reduced lane counts. To understand and model merging behavior, four zones are defined for each type of merging, as shown in figures 1a and 1b.

For on-ramp merging, queuing area (defined as Zone II) may occur along the non-priority minor stream. If traffic condition provided by the priority major stream does not fulfil requirements for the subject vehicle (from minor stream) to enter the merging area, it may queue at the queuing area to wait for a large enough gap to appear. Similarly, for non-priority on-ramp merging, queuing areas may occur for both arriving flows.

2.2. Decision-making of traffic merging

Traffic merging is a complex process of vehicle interactions and it can be considered as a combination of lateral movement and velocity change. Lateral movement can be further divided into two types, lane-changing and lateral drift within the lane (as bounded by delineation markings). Both lane-changing and lateral drift can be caused by two reasons: due to ensuing reduction of lanes or due to obtrusion from a neighbouring vehicle (on same lane or on adjacent lane). Moreover, lateral movements caused by neighbouring vehicles can be further divided as two types, namely to avoid conflicts or to move to a lane with less traffic.

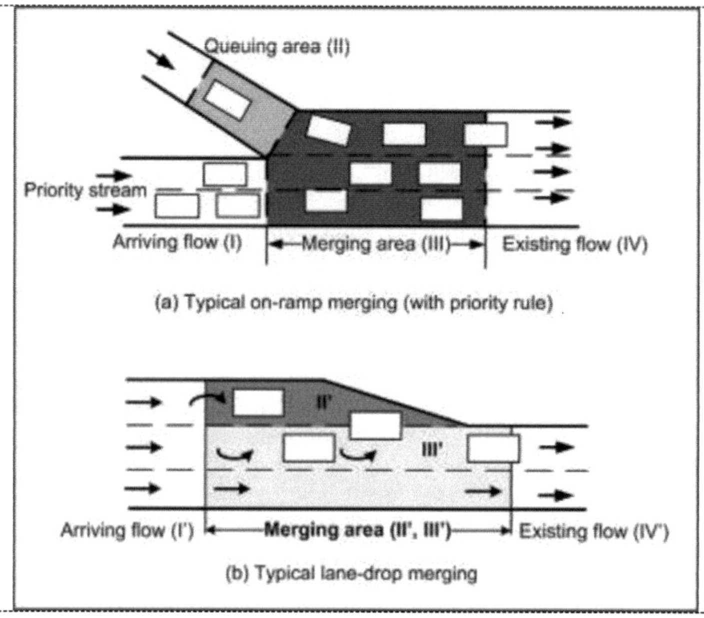

Figure 1. Typical elements of traffic merging.

Interactions at merging traffic involve several steps: Firstly, some vehicles are forced to change lane or drift laterally due to the change of lane configurations. Such lateral movements affect the neighbouring vehicles from both streams and force them to move laterally or change velocity. On the other hand, movement of the subject vehicle is also affected by a neighbouring vehicle.
For each vehicle, the decision-making procedure is described in figure 2. Three groups of input factors, namely lane configuration, operational attributes of neighbouring vehicles from subject stream, and operational attributes of neighbouring vehicles from the other stream, affect decision-making of vehicles. For each merging vehicle, decisions are made at three levels. Firstly, if the subject vehicle does not have priority and is within the queuing area, the first decision is whether to stop or enter the merging area. This decision is made in consideration with operational attributes of vehicles in the other stream. For surrounding vehicles or those vehicles which have decided to enter the merging area, lane change and lateral drift are decided. Last but not least, velocity change for each vehicle is decided.

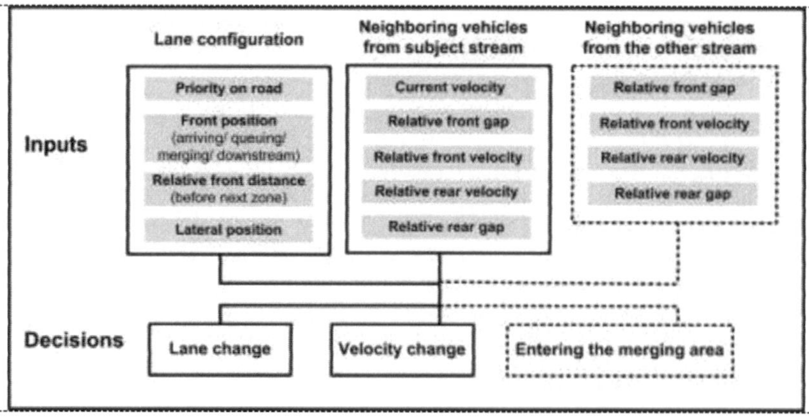

Figure 2. Typical elements of traffic merging (inputs and decisions in dash boxes are applied to vehicles before entering the merging area).

2.3. Fuzzy sets and terms

According to the decision-making process of each individual driver, three fuzzy sets are defined to model vehicle movements, as summarized in Table 1.

Table 1. Linguistic terms for subsets and input factors

Subset	Inputs
Lane configuration (F_{1-1}/ F_{2-1}/ F_{3-1})	**Priority on road** {Yes; No} **Front position** {Arriving; Queuing; Merging; Downstream} **Distance to entering next zone** {Far; Medium; Close} **Lateral position** {Lane 1; Lane 2;…}
Neighbouring vehicles from subject stream (F_{1-2}/ F_{2-2}/ F_{3-2})	**Current velocity** {Slow; Medium; Fast} **Relative front gap** {Close; Medium; Far} **Relative front velocity** {Opening fast; Opening; About zero; Closing; Closing fast} **Relative rear gap** {Close; Medium; Far} **Relative rear velocity** {Opening fast; Opening; About zero; Closing; Closing fast} **Relative lateral gap** {Close; Medium; Far}
Neighbouring vehicles from the other stream (F_4)	**Relative front gap** {Close; Medium; Far} **Relative front velocity** {Opening fast; Opening; About zero; Closing; Closing fast} **Relative rear gap** {Close; Medium; Far} **Relative rear velocity** {Opening fast; Opening; About zero; Closing; Closing fast} **Relative lateral gap** {Close; Medium; Far}

2.4. Proposed FCA approach

In the proposed FCA model, vehicle movements, which are controlled by fuzzy sets and rules, are discretised on a cell space. A smaller cell size is chosen to simulate multiple vehicle types in this study, as shown in figure 3. Each cell is a 0.9 m by 0.9 m square. Lane width is taken to be 3.6 m and therefore four rows of cells constitute one lane. According to physical sizes of the three vehicle types, a car takes 5×2 cells, a heavy vehicle occupies 13×3 cells while a motorcycle takes 3×1 cells. Moreover, for each vehicle, a leading cell (shown as black cell in figure 3) is defined to represent the exact position. The leading cell is also 'aligned' to the driver's position in using Singapore's driving convention.

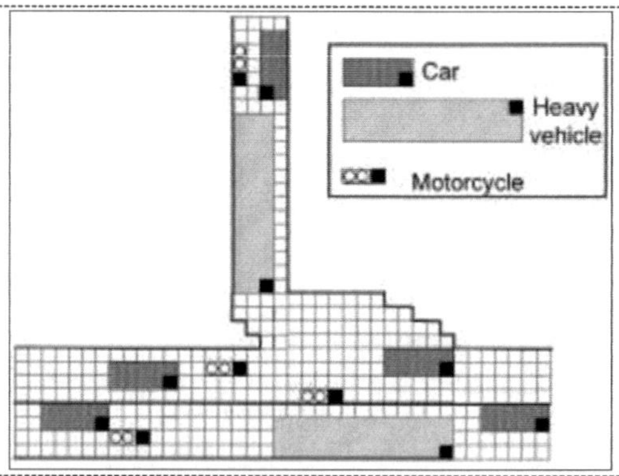

Figure 3. CA model for mixed traffic flow at junction approach.

3. Simulation Experiments and DEA/AHP evaluation

3.1. Methodology

In this study, data envelopment analysis (DEA) and analytic hierarchy process (AHP) are applied to evaluate capacity and safety performance of merging area. The methodology was developed respectively [13,14]. It has been applied for highway geometric design by Chen et al. [12].

Two-stage DEA/AHP model is used to evaluate the different geometric scenarios, as explained in the following. The evaluation attributes include both capacity and safety performance from FCA. These scenarios are then ranked according to the outputs of traffic performance values. The most preferred geometric design is thereby suggested. Four lane-drop distance (A=30m, B=60m, C=90m, D=120m) are tested in four possible traffic conditions with different traffic volume and speed limit. Average simulation results of 10 simulation runs are listed in Table 2.

3.2. Evaluation and analysis

The simulation results of four lane-drop distances under four different traffic conditions generate the pairwise comparison matrices through the DEA method. The comparative efficiencies of four scenarios by AHP ranking are demonstrated in Table 3. The optimal lane-drop distances, which consider both capacity and safety for each traffic condition, are marked in bold.

According to optimal lane-drop distances suggested by DEA/AHP approach, the following findings can be summarised: 1) Effect of traffic volume and speed limit

Both traffic volume and speed limit affects optimal lane-drop distance. A higher traffic volume as well as higher speed limit shall require longer lane-drop distance.

2) Optimal lane-drop distance

In all the four traffic conditions, the maximum optimal lane-drop distance is 90m. The 4th lane-drop distance (D=120m) is not optimal in all the traffic conditions. Therefore, it is suggested that it is not always better when more space is given for merging and an optimal lane-drop distance exists at downstream merging area.

4. Conclusions

This study evaluates microscopic vehicle movements and decision-making process of merging traffic through microscopic simulation based on Fuzzy Logic and Cellular Automata (CA). An improved FCA model is developed by applying fuzzy sets and membership functions to generate driver's responses that are embedded into the CA model. Four different fuzzy sets are designed to simulate driver's response due to lane configuration, surrounding vehicles from the same traffic stream as well as the other, competing traffic stream. Movement characteristics and membership functions are calibrated based on field observations at eight survey sites.

Table 2. Simulation results

Traffic condition	Traffic volume (veh/lane)		Speed limit (km/h)	Lane-drop distance	Travel time (s)	DOC	TTC (s)
	Major	Minor					
1	100	100	60	A	63.1	0.63	2.09
				B	61.7	0.62	2.21
				C	59.3	0.57	2.58
				D	57.2	0.32	2.86
2	100	100	50	A	78.2	0.74	2.01
				B	75.7	0.71	2.54
				C	72.5	0.58	2.74
				D	67.2	0.47	2.82
3	200	100	60	A	69.3	0.81	1.78
				B	67.4	0.77	1.94
				C	66.5	0.69	2.22
				D	59.1	0.65	2.48
4	200	100	50	A	81.4	0.72	1.98
				B	79.3	0.63	2.14
				C	74.2	0.47	2.45
				D	72.9	0.42	2.52

Table 3. Calculated comparative efficiency of land-drop distance scenarios by DEA/AHP

Traffic condition	Lane-drop distance	Travel time (s)	DOC	TTC (s)	Efficiency
1	A	0.2339	0.2044	0.1988	0.8590
	B	0.2416	0.2130	0.2512	0.9090
	C	**0.2523**	**0.2608**	**0.2710**	**1.0000**
	D	0.2722	0.3218	0.2789	0.9830
2	**A**	**0.2387**	**0.1965**	**0.2146**	**1.0000**
	B	0.2441	0.1996	0.2269	0.5287
	C	0.2540	0.2171	0.2649	0.4115
	D	0.2633	0.3868	0.2936	0.4922
3	A	0.2357	0.2236	0.2114	0.9746
	B	0.2423	0.2352	0.2304	0.9566
	C	**0.2456**	**0.2625**	**0.2637**	**1.0000**
	D	0.2764	0.2787	0.2945	0.9830
4	A	0.2358	0.1856	0.2178	0.9984
	B	**0.2421**	**0.2121**	**0.2354**	**1.0000**
	C	0.2587	0.2843	0.2695	0.9782
	D	0.2633	0.3181	0.2772	0.9831

The results are then evaluated through the data envelopment analysis (DEA) and analytic hierarchy process (AHP). Preferable geometric layout and control strategies are thereby suggested for various traffic conditions. It is found that an optimal lane-drop distance exists at downstream merging area and it is affected by both traffic volume and speed limit.

Although the CA model is widely used in simulating road traffic, the major application is on capacity assessment due to its computational efficiency. However, with improvements involving the decision-making process, the FCA model is more able to estimate dynamic driver's responses due to different lane configurations and surrounding vehicles. Simulation experiments demonstrate that this approach is able to assess lane-changing and merging behaviour.

Some improvements can be made to refine and extend the applications of proposed FCA model in future works. For example, since the proposed FCA model involves cognition system, field observations alone are not sufficient to calibrate or justify the model. In future study, driving experiments (based on real driving or simulator) and questionnaire surveys will be conducted to better understand behaviour, perception and attitude of road users.

5. References

[1] Jin, W. L. Transportation Research Part B 44, 1084-1103 (2010)
[2] Lee, J. J. North Carolina State University, Ph.D. Thesis (2005)
[3] Sarvi, M. and Kuwahara, M. IEEE Transactions on Intelligent Transportation Systems 8 (3), 470-479 (2007)
[4] Xiao, L. L., 2014. Transportation Research Part E 64, 48-70 (2014)
[5] Hidas, P. Transportation Research Part C 13 (1), 37-62 (2005)
[6] Ni, D. and Leonard, J., Applied Mathematical Modelling 29 (11), 1054-1072 (2005)
[7] Carlson, R. C., Ragias, A., Papamichail, I. and Papageorgiou, M. 19th Mediterranean Conference on Control and Automation Greece (2011)
[8] St-Aubin, P., Miranda-Moreno, L., Saunier, N. Transportation Research Part C: Emerging Techniques 36, 284-295 (2013)
[9] Chiou, Y. C. Huang, Y. F. Journal of Advanced Transportation 47(1), 43-60 (2003)
[10] Placzek, B. Online resource in Cornell University Library. http://arxiv.org/abs/1112.4631 (2012)
[11] Chai, C., Wong, Y. D. Computer-Aided Civil and Infrastructure Engineering 30 (12), 951-964 (2015)
[12] Chen, S. K., Mao, B. H., Liu, S., Sun, Q. X., Wei, W., Zhan, L. X. Transportation Research Part C: Emerging Techniques 36, 381-393 (2013)
[13] Charnes, A., Cooper, W.W., Rhodes, E. European Journal of Operational Research 2 (6), 429-444 (1978)
[14] Saaty, T. L. Multi Criteria Decision Making: The Analytic Hierarchy Process. McGraw-Hill, New York (1980)

Acknowledgments
This work is supported by Singapore Ministry of Education Academic Research Fund Tier 2 MOE 2013-T2-2-073.

Some results on circulant and skew circulant type matrices with k-Fibonacci sequences

Ying Zhang

Department of Mathematics, Dalian Maritime University, Dalian 116024, China

zhgyg77@sina.com

Abstract. In this paper, we consider the circulant matrices, skew circulant matrices and skew left circulant matrices with k-Fibonacci sequences. We compute their eigenvalues by diagonalizing these matrices. Furthermore, the maximum column sum matrix norms, the maximum row sum matrix norms, the Frobenius norms, the spectral norms, and the bounds for the spread of these matrices are given with the properties of k-Fibonacci sequences.

1. Introduction

Circulant matrices and their generalization have a wide range of applications in signal coding theory, image processing, digital image disposal, self-regress design and so on [1][2][3][4][5]. Specifically, the computing of the eigenvalues, norms and bounds of spread of the nonsingular circulant matrices plays a basic role in the above applications. Though various computation algorithms have been reported [6], the computation complexity of these algorithms becomes rather high with the increasing of the order of matrices.

Jaiswal evaluated some determinants of circulant matrices whose elements are the generalized Fibonacci numbers [7]. Subsequently, Lind computed the determinants of circulant and skew-circulant matrices involving Fibonacci numbers [8]. These works aroused promptly wide public concern, and people began researching into circulant type matrices with special sequences [9][10][11].

Recently, Falcón proposed *k*-Fibonacci sequence and obtained Binet's formulas [12][13]. For any positive real number k and $n \geq 1$, the *k*-Fibonacci sequences $\{F_{k,n}\}_{n\in N}$ are defined recurrently by $F_{k,n+1} = kF_{k,n} + F_{k,n-1}$, where $F_{k,0} = 0$ and $F_{k,1} = 1$. In particular, $\{F_{1,n}\}_{n\in N}$ is the classical Fibonacci sequence, $\{F_{2,n}\}_{n\in N}$ is the classical Pell sequence, and $\{F_{3,n}\}_{n\in N}$ is the product of Fibonacci and Lucas numbers. Thus, it is very interesting to study the explicit formulae for the eigenvalues of the circulant and skew circulant type matrices with k-Fibonacci sequences, which include many existing sequences as special cases.

The main aim of of this paper is to establish some useful formulae for the eigenvalues of circulant, skew circulant and skew left circulant matrices whose elements are the *k*-Fibonacci sequences. Using the properties of *k*-Fibonacci sequences [14][15], the maximum column sum matrix norms, the maximum row sum matrix norms, the Frobenius norms, the spectral norms and bounds for the spread of these matrices will also be given. In particular, we will make the following contributions:

- We obtain an *explicit formula* for spectral norms of circulant matrices with *k*-Fibonacci sequence. We mention a related work in [16], where *the upper and lower bounds* for the spectral norms of circulant matrices with (*k*,*h*)-Fibonacci numbers are discussed.

- The spectral norms formulae in this paper can be valid for any order skew circulant type matrices.

2. Prelimimaries

Lemma 2.1. (Binet's formula [12]) The nth k-Fibonacci number is given by

$$F_{k,n} = \frac{r_1^n - r_2^n}{r_1 - r_2}, \tag{1}$$

where r_1, r_2 are the roots of the characteristic equation $r^2 - kr - 1 = 0$, and $r_1 > r_2$.

Lemma 2.2. (see [13]) Let $S_{k,n}$ be the sum of the first $n+1$ terms of the k-Fibonacci sequence, that is $S_{k,n} = \sum_{j=0}^{n} F_{k,j}$ .Then

$$\text{(i) } S_{k,n} = \frac{1}{k}(F_{k,n+1} + F_{k,n}) - \frac{1}{k}; \quad \text{(ii) } \sum_{j=1}^{n} F_{k,2j} = \frac{1}{k}(F_{k,2n+1} - 1); \quad \text{(iii) } \sum_{j=0}^{n} F_{k,2j+1} = \frac{1}{k} F_{k,2n+2}. \tag{2}$$

Lemma 2.3. (see [15]) Let $\{F_{k,n}\}$ be k-Fibonacci sequence, then

$$\sum_{j=0}^{n} F_{k,j}^2 = \frac{1}{k} F_{k,n} F_{k,n+1} \tag{3}$$

Lemma 2.4. Let $\{F_{k,n}\}$ be k-Fibonacci sequence, then

$$\sum_{j=0}^{n} jF_{k,j} = \frac{2}{k^2} + \frac{kn+k-2}{k^2} F_{k,n} + \frac{kn-2}{k^2} F_{k,n+1}. \tag{4}$$

Proof.

$$\sum_{j=0}^{n} jF_{k,j} = \sum_{j=0}^{n} j \cdot \frac{r_1^j - r_2^j}{r_1 - r_2} = \frac{1}{r_1 - r_2}\left[\frac{\frac{r_1^2(1-r_1^{n-1})}{1-r_1} + 1 - nr_1^{n+1}}{1-r_1} - \frac{\frac{r_2^2(1-r_2^{n-1})}{1-r_2} + 1 - nr_2^{n+1}}{1-r_2}\right]$$

$$= \frac{1}{k^2}[2 - F_{k,n-1} + (kn-2)F_{k,n} + (kn-1)F_{k,n+1}] = \frac{2}{k^2} + \frac{kn+k-2}{k^2} F_{k,n} + \frac{kn-2}{k^2} F_{k,n+1} \tag{5}$$

3. Circulant matrix with the k-Fibonacci sequence

The $n \times n$ circulant matrix $C_n := Circ(c_1, \cdots, c_n)$ is defined as

$$C_n := \begin{pmatrix} c_1 & c_2 & \cdots & c_n \\ c_n & c_1 & \cdots & c_{n-1} \\ \vdots & \vdots & \ddots & \vdots \\ c_2 & c_3 & \cdots & c_1 \end{pmatrix}. \tag{6}$$

Let $\mathbf{F}_{k,n} := Circ(F_{k,1}, F_{k,2}, \cdots, F_{k,n})$ be the circulant matrices associated with the k-Fibonacci sequences. In this section, we obtain an expression for eigenvalues of $\mathbf{F}_{k,n}$ and compute four kinds of norms and bounds for the spread of $\mathbf{F}_{k,n}$.

Theorem 3.1. Let $\mathbf{F}_{k,n} := Circ(F_{k,1}, F_{k,2}, \cdots, F_{k,n})$ be a circulant matrix, then the eigenvalues of $\mathbf{F}_{k,n}$ are

$$\frac{1 - F_{k,n+1} - F_{k,n} e^{i\frac{2\pi(j-1)}{n}}}{1 - ke^{i\frac{2\pi(j-1)}{n}} - e^{i\frac{4\pi(j-1)}{n}}}, (j = 1, \cdots, n). \tag{7}$$

Proof. By Theorem 7 in [17], let $\chi(x)=\sum_{l=1}^{n}F_{k,l}x^{l-1}$ and $\theta_j=e^{i\frac{2\pi(j-1)}{n}}$ $(j=1,\cdots,n)$, we know the eigenvalues of $\mathbf{F}_{k,n}$ are

$$\chi(\theta_j)=\sum_{l=1}^{n}F_{k,l}e^{i\frac{2\pi(j-1)(l-1)}{n}}$$

$$=\frac{1}{r_1-r_2}\sum_{l=1}^{n}(r_1^l-r_2^l)e^{i\frac{2\pi(j-1)(l-1)}{n}}=\frac{1}{r_1-r_2}[\frac{r_1(1-r_1^n)}{1-r_1e^{i\frac{2\pi(j-1)}{n}}}-\frac{r_2(1-r_2^n)}{1-r_2e^{i\frac{2\pi(j-1)}{n}}}]$$

$$=\frac{1}{r_1-r_2}[\frac{r_1-r_2-(r_1^{n+1}-r_2^{n+1})-(r_1^n-r_2^n)e^{i\frac{2\pi(j-1)}{n}}}{1-(r_1+r_2)e^{i\frac{2\pi(j-1)}{n}}+r_1r_2e^{i\frac{4\pi(j-1)}{n}}}]=\frac{1-F_{k,n+1}-F_{k,n}e^{i\frac{2\pi(j-1)}{n}}}{1-ke^{i\frac{2\pi(j-1)}{n}}-e^{i\frac{4\pi(j-1)}{n}}} \quad (8)$$

Theorem 3.2. Let $\mathrm{F}_{k,n}:=Circ(F_{k,1},F_{k,2},\cdots,F_{k,n})$ be a circulant matrix, then the maximum column sum matrix norm, the maximum row sum matrix norm, the Frobenius norm, and the spectral norm of $\mathbf{F}_{k,n}$ are respectively given as follows

$$\|\mathbf{F}_{k,n}\|_1=\|\mathbf{F}_{k,n}\|_\infty=\frac{1}{k}(F_{k,n+1}+F_{k,n})-\frac{1}{k}, \quad (9)$$

$$\|\mathbf{F}_{k,n}\|_F=\sqrt{\frac{n}{k}F_{k,n}F_{k,n+1}}, \quad (10)$$

$$\|\mathbf{F}_{k,n}\|_2=\frac{1}{k}(F_{k,n+1}+F_{k,n})-\frac{1}{k}. \quad (11)$$

Proof. By (2), we have

$$\|\mathbf{F}_{k,n}\|_1=\|\mathbf{F}_{k,n}\|_\infty=\sum_{j=1}^{n}F_{k,j}=\frac{1}{k}(F_{k,n+1}+F_{k,n})-\frac{1}{k}. \quad (12)$$

Furthermore, by (3) we know

$$(\|\mathbf{F}_{k,n}\|_F)^2=n\sum_{j=1}^{n}F_{k,j}^2=\frac{n}{k}F_{k,n}F_{k,n+1}. \quad (13)$$

According to Theorem 1 in [18], since $\mathbf{F}_{k,n}$ is normal, irreducible and entrywise nonnegative, then

$$\|\mathbf{F}_{k,n}\|_2=\sum_{j=1}^{n}F_{k,j}=\frac{1}{k}(F_{k,n+1}+F_{k,n})-\frac{1}{k}.$$

Theorem 3.3. Let $\mathrm{F}_{k,n}:=Circ(F_{k,1},F_{k,2},\cdots,F_{k,n})$ be a circulant matrix, and $s(\mathbf{F}_{k,n})$ denote the spread of $\mathbf{F}_{k,n}$. Then we have

$$s(\mathbf{F}_{k,n})\le\sqrt{2n(\frac{1}{k}F_{k,n}F_{k,n+1}-1)}, \quad (14)$$

$$s(\mathbf{F}_{k,n})\ge\frac{n}{n-1}[\frac{1}{k}(F_{k,n+1}+F_{k,n})-\frac{1}{k}-1]. \quad (15)$$

Proof. Let $tr(\mathbf{F}_{k,n})$ denote the trace of $\mathbf{F}_{k,n}$, then $tr(\mathbf{F}_{k,n})=nF_{k,1}$. By (10), we know

$$s(\mathbf{F}_{k,n})\le\sqrt{2\|\mathrm{F}_{k,n}\|_F^2-\frac{2}{n}|tr(\mathrm{F}_{k,n})|^2}\le\sqrt{2n(\frac{1}{k}\mathrm{F}_{k,n}\mathrm{F}_{k,n+1}-1)}. \quad (16)$$

Since

$$\sum_{s\neq t} a_{st} = n[\frac{1}{k}(F_{k,n+1}+F_{k,n})-\frac{1}{k}-1], \tag{17}$$

then, we have

$$s(\mathbf{F}_{k,n}) \geq \frac{1}{n-1}|\sum_{s\neq t} a_{st}| = \frac{n}{n-1}[\frac{1}{k}(F_{k,n+1}+F_{k,n})-\frac{1}{k}-1].$$

4. Skew circulant matrix with the k-Fibonacci sequence

The $n\times n$ skew circulant matrix $SC_n := SCirc(c_1,\cdots, c_n)$ is define as

$$SC_n := \begin{pmatrix} c_1 & c_2 & \cdots & c_n \\ -c_n & c_1 & \cdots & c_{n-1} \\ -c_{n-1} & c_n & \cdots & c_{n-2} \\ \vdots & \vdots & \ddots & \vdots \\ -c_2 & -c_3 & \cdots & c_1 \end{pmatrix}. \tag{19}$$

Let $\mathbf{SF}_{k,n} := SCirc(F_{k,1}, F_{k,2},\cdots, F_{k,n})$ be the skew circulant matrices associated with the k-Fibonacci sequences. In this section, we obtain the expression for eigenvalues of $\mathbf{SF}_{k,n}$, and compute four kinds of norms and bounds for the spread of $\mathbf{SF}_{k,n}$.

Theorem 4.1. Let $\mathbf{SF}_{k,n} := SCirc(F_{k,1}, F_{k,2},\cdots, F_{k,n})$ be a skew circulant matrix, then the eigenvalues of $\mathbf{SF}_{k,n}$ are

$$\frac{1+F_{k,n+1}+F_{k,n}e^{i\frac{\pi+2\pi(j-1)}{n}}}{1-ke^{i\frac{\pi+2\pi(j-1)}{n}}-e^{i\frac{2\pi+4\pi(j-1)}{n}}}, (j=1,\cdots,n). \tag{20}$$

Proof. Let $\chi(x)=\sum_{l=1}^{n} F_{k,l}x^{l-1}$ and

$$\vartheta_j = e^{i\frac{\pi+2\pi(j-1)}{n}} \quad (j=1,\cdots,n). \tag{21}$$

Then, by Theorem 7 in [17], we can compute the eigenvalues of $\mathbf{SF}_{k,n}$ as follows

$$\begin{aligned}\chi(\vartheta_j) &= \sum_{l=1}^{n} F_{k,l} e^{i\frac{[\pi+2\pi(j-1)](l-1)}{n}} \\ &= \frac{1}{r_1-r_2}\sum_{l=1}^{n}(r_1^l - r_2^l)e^{i\frac{[\pi+2\pi(j-1)](l-1)}{n}} = \frac{1}{r_1-r_2}[\frac{r_1(1+r_1^n)}{1-r_1e^{i\frac{\pi+2\pi(j-1)}{n}}} - \frac{r_2(1+r_2^n)}{1-r_2e^{i\frac{\pi+2\pi(j-1)}{n}}}] \\ &= \frac{1}{r_1-r_2}[\frac{r_1-r_2+(r_1^{n+1}-r_2^{n+1})+(r_1^n-r_2^n)e^{i\frac{\pi+2\pi(j-1)}{n}}}{1-(r_1+r_2)e^{i\frac{\pi+2\pi(j-1)}{n}}+r_1r_2e^{i\frac{2\pi+4\pi(j-1)}{n}}}] = \frac{1+F_{k,n+1}+F_{k,n}e^{i\frac{\pi+2\pi(j-1)}{n}}}{1-ke^{i\frac{\pi+2\pi(j-1)}{n}}-e^{i\frac{2\pi+4\pi(j-1)}{n}}}.\end{aligned} \tag{22}$$

For convenience of presentation, we introduce some notations. We use A^T, A^*, A^H to denote the transpose, the conjugate, and the conjugate transpose of A, respectively. We also use $\lambda_{\max}(A)$ to represent the maximum eigenvalue of A.

Theorem 4.2. Let $\mathbf{SF}_{k,n} := SCirc(F_{k,1}, F_{k,2},\cdots, F_{k,n})$ be a skew circulant matrix, then the maximum column sum matrix norm, the maximum row sum matrix norm, the Frobenius norm, and the spectral norm of $\mathbf{SF}_{k,n}$ are respectively given as follows

$$\|\mathbf{SF}_{k,n}\|_1 = \|\mathbf{SF}_{k,n}\|_\infty = \frac{1}{k}(F_{k,n+1}+F_{k,n})-\frac{1}{k}, \tag{23}$$

$$\| \mathbf{SF}_{k,n} \|_F = \sqrt{\frac{n}{k} F_{k,n} F_{k,n+1}}, \tag{24}$$

$$\| \mathbf{SF}_{k,n} \|_2 = \max_{1 \le j \le n} \sqrt{\chi(\vartheta_j)\chi^*(\vartheta_j)}, \tag{25}$$

where $\chi(\vartheta_j)(j=1,\cdots,n)$ is defined by (22).

Proof. The deductions of (23) and (24) essentially are the same as Theorem 3.3, thus the details are omitted here. In the following, we give the proof of (25).

Write

$$G = \begin{pmatrix} 0 & 1 & 0 & \cdots & 0 \\ 0 & 0 & 1 & \cdots & 0 \\ \vdots & \vdots & \vdots & \ddots & \vdots \\ 0 & 0 & 0 & \cdots & 1 \\ -1 & 0 & 0 & \cdots & 0 \end{pmatrix}, \tag{26}$$

then $SCirc(c_1,\cdots,c_n) = \sum_{j=1}^{n} c_j G^{j-1}$. Noticing $G^T G = G G^T$, we have that

$$SC_n \cdot SC_n^T = (\sum_{j=1}^{n} c_j G^{j-1})(\sum_{j=1}^{n} c_j (G^{j-1})^T) = (\sum_{j=1}^{n} c_j (G^{j-1})^T)(\sum_{j=1}^{n} c_j G^{j-1}) = SC_n^T \cdot SC_n \tag{27}$$

which shows SC_n is normal matrix. Thus $\mathbf{SF}_{k,n}$ is normal matrix. Therefore, there exists unitary matrix Q such that

$$Q^H \cdot \mathbf{SF}_{k,n} \cdot Q = diag(\chi(\vartheta_1),\cdots,\chi(\vartheta_n)), \tag{28}$$

$$Q^H \cdot \mathbf{SF}_{k,n}^T \cdot Q = diag(\chi^*(\vartheta_1),\cdots,\chi^*(\vartheta_n)). \tag{29}$$

Multiplying the above two equations, we have

$$Q^H \cdot \mathbf{SF}_{k,n}^T \cdot \mathbf{SF}_{k,n} \cdot Q = diag(\chi(\vartheta_1)\chi^*(\vartheta_1),\cdots,\chi(\vartheta_n)\chi^*(\vartheta_n)). \tag{30}$$

Using (30), we obtain that

$$\| \mathbf{SF}_{k,n} \|_2 = \sqrt{\lambda_{\max}(\mathbf{SF}_{k,n}^T \cdot \mathbf{SF}_{k,n})} \tag{31}$$

$$= \max_{1 \le j \le n} \sqrt{\chi(\vartheta_j)\chi^*(\vartheta_j)} \tag{32}$$

Theorem 4.3. Let $\mathbf{SF}_{k,n} := SCirc(F_{k,1}, F_{k,2},\cdots,F_{k,n})$ be a skew circulant matrix, then we have the following bounds for the spectral of $\mathbf{SF}_{k,n}$

$$s(\mathbf{SF}_{k,n}) \le \sqrt{2n(\frac{1}{k} F_{k,n} F_{k,n+1} - 1)}, \tag{33}$$

$$\boldsymbol{s(\mathbf{SF}_{k,n}) \ge \frac{1}{n-1}[(\frac{4}{k^2} - \frac{n-2}{k})F_{k,n+1} + (\frac{4}{k^2} - \frac{n}{k})F_{k,n} - (\frac{4}{k^2} + \frac{n+2}{k} + n)].} \tag{34}$$

Proof. It is obvious that

$$s(\mathbf{SF}_{k,n}) \le \sqrt{2n(\frac{1}{k} F_{k,n} F_{k,n+1} - 1)}. \tag{35}$$

By (4), we have

$$\sum_{s \ne t} a_{st} = \sum_{l=2}^{n} (n-(l-1))F_{k,l} - \sum_{l=2}^{n} (l-1)F_{k,l} = (n+2)\sum_{l=2}^{n} F_{k,l} - 2\sum_{l=2}^{n} l F_{k,l}$$

$$= (n+2)[\frac{1}{k}(F_{k,n+1} + F_{k,n}) - \frac{1}{k} - 1] - 2(\frac{2}{k^2} + \frac{kn+k-2}{k^2} F_{k,n} + \frac{kn-2}{k^2} F_{k,n+1} - 1)$$

$$= (\frac{4}{k^2} - \frac{n-2}{k})F_{k,n+1} + (\frac{4}{k^2} - \frac{n}{k})F_{k,n} + (\frac{4}{k^2} + \frac{n+2}{k} + n).$$

Hence, we get

$$s(\mathbf{SF}_{k,n}) \geq \frac{1}{n-1}[(\frac{4}{k^2}-\frac{n-2}{k})F_{k,n+1}+(\frac{4}{k^2}-\frac{n}{k})F_{k,n}-(\frac{4}{k^2}+\frac{n+2}{k}+n)].$$

5. Skew left circulant matrix with the k-Fibonacci sequence

The $n \times n$ skew left circulant matrix $SLC_n := SLCirc(c_1, \cdots, c_n)$ is define as

$$SLC_n := \begin{pmatrix} c_1 & c_2 & \cdots & c_n \\ c_2 & c_3 & \cdots & -c_1 \\ c_3 & c_4 & \cdots & -c_2 \\ \vdots & \vdots & \ddots & \vdots \\ c_n & -c_1 & \cdots & -c_{n-1} \end{pmatrix}. \tag{37}$$

Let $\mathbf{SLF}_{k,n} := SLCirc(F_{k,1}, \cdots, F_{k,n})$ be the skew left circulant matrices associated with k-Fibonacci sequences. In this section, we obtain the expression for eigenvalues of $\mathbf{SLF}_{k,n}$, and compute four kinds of norms and bounds for the spread of $\mathbf{SLF}_{k,n}$.

Theorem 5.1. Let $\mathbf{SLF}_{k,n} := SLCirc(F_{k,1}, F_{k,2}, \cdots, F_{k,n})$ be a skew left circulant matrix, then the eigenvalues of $\mathbf{SLF}_{k,n}$ are

$$\begin{cases} \pm\sqrt{\varphi(\vartheta_t)\psi(\vartheta_t)}(t=1,\cdots,\frac{n}{2}), & if\ n\ is\ even; \\ \pm\sqrt{\varphi(\vartheta_t)\psi(\vartheta_t)}(t=1,\cdots,\frac{n-1}{2}) and\ \frac{1}{k}(F_{k,n+1}-F_{k,n}+1) & if\ n\ is\ odd. \end{cases} \tag{38}$$

Here $\vartheta_t(t=1,\cdots,[\frac{n}{2}])$, $\varphi(x)$ and $\psi(x)$ are defined by (21), (43) and (45).

Proof. Let $J_n = (e_n, e_{n-1}, \cdots, e_1)$ stand for the $n \times n$ reverse unit matrix (e_i denotes the ith column of $n \times n$ unit matrix). Note that,

$$J_n \cdot SLCirc(F_{k,1}, \cdots, F_{k,n}) = SCirc(F_{k,n}, -F_{k,1}, \cdots, -F_{k,n-1}) \tag{39}$$

and

$$SLCirc(F_{k,1}, \cdots, F_{k,n}) \cdot J_n = SCirc(F_{k,n}, F_{k,n-1}, \cdots, F_{k,1}). \tag{40}$$

By Theorem 7 in [17], there exists

$$V = \begin{pmatrix} 1 & 1 & 1 & \cdots & 1 \\ \vartheta_1 & \vartheta_2 & \vartheta_3 & \cdots & \vartheta_n \\ \vartheta_1^2 & \vartheta_2^2 & \vartheta_3^2 & \ddots & \vartheta_n^2 \\ \vdots & \vdots & \vdots & \ddots & \vdots \\ \vartheta_1^{n-1} & \vartheta_2^{n-1} & \vartheta_3^{n-1} & \cdots & \vartheta_n^{n-1} \end{pmatrix}, \tag{41}$$

such that

$$SCirc(F_{k,n}, -F_{k,1}, \cdots, -F_{k,n-1}) = V diag(\varphi(\vartheta_1), \cdots, \varphi(\vartheta_n))V^{-1}, \tag{42}$$

where

$$\varphi(x) = F_{k,n} - F_{k,1}x - F_{k,2}x^2 - \cdots - F_{k,n-1}x^{n-1}, \tag{43}$$

and

$$SCirc(F_{k,n}, F_{k,n-1}, \cdots, F_{k,1}) = V diag(\psi(\vartheta_1), \cdots, \psi(\vartheta_n))V^{-1}. \tag{44}$$

Here

$$\psi(x) = F_{k,n} + F_{k,n-1}x + F_{k,n-2}x^2 + \cdots + F_{k,1}x^{n-1}. \tag{45}$$

Using (39), (40), (42) and (44), we have that

$$J_n \cdot SLCirc(F_{k,1}, \cdots, F_{k,n})^2 J_n = V diag(\varphi(\vartheta_1)\psi(\vartheta_1), \cdots, \varphi(\vartheta_n)\psi(\vartheta_n))V^{-1}. \tag{46}$$

For $t=1,\cdots,[\frac{n}{2}]$, where $[x]$ denotes the maximum integer is less than x, there hold

$$\varphi(\vartheta_t)=F_{k,n}-\sum_{u=1}^{n-1}F_{k,u}\vartheta_t^u, \tag{47}$$

and

$$\varphi(\vartheta_{n+1-t})=F_{k,n}+\sum_{u=n-1}^{1}F_{k,u}\vartheta_{n+1-t}^{n-u}. \tag{48}$$

Considering that

$$\vartheta_t\vartheta_{n+1-t}=1 \ and \ \vartheta_t^n=\vartheta_{n+1-t}^n=-1, \tag{49}$$

We have

$$\vartheta_{n+1-t}^{n-u}=-(\vartheta_{n+1-t})^{-u}=-\vartheta_t^u. \tag{50}$$

Combining (47),(48) and (50), we have

$$\varphi(\vartheta_t)=\psi(\vartheta_{n+1-t}), \tag{51}$$

In a similar manner, we have $\varphi(\vartheta_{n+1-t})=\psi(\vartheta_t)$. Thus,

$$\varphi(\vartheta_t)\psi(\vartheta_t)=\varphi(\vartheta_{n+1-t})\psi(\vartheta_{n+1-t}). \tag{52}$$

This means that if n is even, the eigenvalues of $\mathbf{SLF}_{k,n}$ are

$$\pm\sqrt{\varphi(\vartheta_t)\psi(\vartheta_t)},(t=1,\cdots,\frac{n}{2}).$$

If n is odd, the eigenvalues of $\mathbf{SLF}_{k,n}$ are still

$$\pm\sqrt{\varphi(\vartheta_t)\psi(\vartheta_t)},(t=1,\cdots,\frac{n-1}{2}),$$

and the eigenvalues left is equal to the trace of $\mathbf{SLF}_{k,n}$, namely $\frac{1}{k}(F_{k,n+1}-F_{k,n}+1)$. Now we obtain (38).

Theorem 5.2. Let $\mathbf{SLF}_{k,n}:=SLCirc(F_{k,1},F_{k,2},\cdots,F_{k,n})$ be a skew left circulant matrix, then the maximum column sum matrix norm, the maximum row sum matrix norm, the Frobenius norm, and the spectral norm of $\mathbf{SLF}_{k,n}$ are given by

$$\|\mathbf{SLF}_{k,n}\|_1=\|\mathbf{SLF}_{k,n}\|_\infty=\frac{1}{k}(F_{k,n+1}+F_{k,n})-\frac{1}{k}, \tag{53}$$

$$\|\mathbf{SLF}_{k,n}\|_F=\sqrt{\frac{n}{k}F_{k,n}F_{k,n+1}}, \tag{54}$$

$$\|\mathbf{SLF}_{k,n}\|_2=\begin{cases}\max\limits_{1\le t\le\frac{n}{2}}\sqrt{\varphi(\vartheta_t)\psi(\vartheta_t)}, & if\ n\ is\ even;\\ \max\{\max\limits_{1\le t\le\frac{n-1}{2}}\sqrt{\varphi(\vartheta_t)\psi(\vartheta_t)},\frac{1}{k}(F_{k,n+1}-F_{k,n}+1)\}, & if\ n\ is\ odd.\end{cases} \tag{55}$$

Here $\vartheta_t(t=1,\cdots,[\frac{n}{2}]),\varphi(x)$ and $\psi(x)$ are defined by (21), (43) and (45).

Proof. The procedure for deducing (53) and (54) is essentially the same as in Theorem 3.3, and the proof is omitted here. It is obvious that $\mathbf{SLF}_{k,n}$ is a symmetry matrix, hence $\mathbf{SLF}_{k,n}$ is a normal matrix. Then

$$\|\mathbf{SLF}_{k,n}\|_2=\sqrt{\lambda_{\max}(\mathbf{SLF}_{k,n}^T\cdot\mathbf{SLF}_{k,n})}=\sqrt{\lambda_{\max}(\mathbf{SLF}_{k,n}^2)}. \tag{56}$$

By (38), if n is even, we have

$$\|\mathbf{SLF}_{k,n}\|_2=\max_{1\le t\le\frac{n}{2}}\sqrt{\varphi(\vartheta_t)\psi(\vartheta_t)}.$$

If n is odd, we have

$$\|\mathbf{SLF}_{k,n}\|_2=\max\{\max_{1\le t\le\frac{n-1}{2}}\sqrt{\varphi(\vartheta_t)\psi(\vartheta_t)},\frac{1}{k}(F_{k,n+1}-F_{k,n}+1)\}.$$

Theorem 5.3. Let $\mathbf{SLF}_{k,n}:=SLCirc(F_{k,1},F_{k,2},\cdots,F_{k,n})$ be a skew left circulant matrix, then the bounds for the spread of $\mathbf{SLF}_{k,n}$ are

$$\begin{cases} 2F_{k,n}\le s(\mathbf{SLF}_{k,n})\le\sqrt{\frac{2n}{k}F_{k,n}F_{k,n+1}}, & \text{if } n \text{ is even,}\\ 2F_{k,n}\le s(\mathbf{SLF}_{k,n})\le\sqrt{\frac{2n}{k}F_{k,n}F_{k,n+1}-\frac{2}{nk^2}(F_{k,n+1}-F_{k,n}+1)^2}, & \text{if } n \text{ is odd.}\end{cases}$$

Proof. If n is even, $tr(\mathbf{SLF}_{k,n})=0$, then we have

$$s(\mathbf{SLF}_{k,n})\le\sqrt{\frac{2n}{k}F_{k,n}F_{k,n+1}}. \quad (58)$$

If n is odd, by (2) we have

$$tr(\mathbf{SLF}_{k,n})=\sum_{j=0}^{\frac{n-1}{2}}F_{k,2j+1}-\sum_{j=1}^{\frac{n-1}{2}}F_{k,2j}=\frac{1}{k}(F_{k,n+1}-F_{k,n}+1).$$

Thus

$$s(\mathbf{SLF}_{k,n})\le\sqrt{\frac{2n}{k}F_{k,n}F_{k,n+1}-\frac{2}{nk^2}(F_{k,n+1}-F_{k,n}+1)^2}.$$

In addition, since $\mathbf{SLF}_{k,n}$ is a symmetric matrix, then we have

$$s(\mathbf{SLF}_{k,n})\ge 2\max_{s\ne t}|a_{st}|=2F_{k,n}.$$

6. Conclusion

Circulant, skew circulant and skew left circulant matrices play an important role in engineering application. In this paper, circulant, skew circulant and skew left circulant matrices with k-Fibonacci sequences are considered. Firstly, eigenvalues of these matrix are discussed and the explicit eigenvalue formulae by diagonalizing these matrices are presented. Furthermore, four kinds of norms including the maximum column sum matrix norms, the maximum row sum matrix norms, the Frobenius norms, the spectral norms, and the bounds for the spread of these matrices are also studied. Our work will be useful complement to the existing researches on circulant and skew circulant type matrices.

7. References

[1] Zhao W 2009 The inverse problem of anti-circulant matrices in signal processing *Proc. Paci.-Asia Conf. on Know. Engi. and Soft. Engi.* (Shen-zhen/ China) pp 47-50
[2] Zhao G 2009 The improved nonsingularity on the r-circulant matrices in signal processing *Proc. Int. Conf. on Comp. Tech. and Deve.* (Kota Kinabalu/Malaysia) pp 564-567
[3] Liu V C and Vaidyanathan P P 1988 Circulant and skew-circulant matrices as new normal-form realization of IIR digital fiters *IEEE Transactions on Circuits and Systems* 35(6) pp 625-635
[4] Mayer A, Castiaux A and Vigneron J P 1998 Electronic Green scattering with n-fold symmetry axis from block circulant matrices *Comp. Phys. Comm.* 109(1) pp 81-89
[5] Ng M K 2003 Circulant and skew-circulant splitting methods for Toeplitz systems *Jour. of Comp. and Appl. Math.* 159(1) pp 101-108
[6] Davis P J 1979 *Circulant matrices* (John Wiley Sons/ New York)
[7] Jaiswal D V 1969 On determinants involving generalized Fibonacci numbers *Fibo. Quart* 7 pp 319-330
[8] Lind D A 1970 A Fibonacci circulant *Fibo. Quart* 8 pp 449-455

[9] Yazlik Y and Taskara N 2013 On the inverse of circulant matrix via generalized k-Horadam numbers *Appl. Math. and Comp.* 223 pp 191-196
[10] Bozkurt D and Tam T Y 2012 Determinants and inverses of circulant matrices with Jacobsthal and Jacobsthal-Lucas Numbers *Appl. Math. and Comp.* 219(2) pp 544-551
[11] Bueno A C F 2013 Left circulant matrices with arithmetic sequence *South Asian Jour. of Math.* 3(5) pp 322-325
[12] Sergio F and A Plaza 2007 The k-Fibonacci sequence and the Pascal 2-triangle *Chao. Soli. and Frac.* 33(1) 38-49
[13] Sergio F and Plaza A 2007 On the Fibonacci k-numbers *Chao. Soli. and Frac.* 32(5) pp 1615-1624
[14] Bolat C 2010 On the properties of k-Fibonacci numbers *Int. J. Cont. Math. Scie.* 5(22) pp 1097-1105
[15] Kilic E 2008 Sums of the squares of terms of sequence {un} *Indi. Acad. Sci. (Math. Sci.)* 118(1) pp 27-41
[16] Shen S 2010 On the spectral norms of r-circulant matrices with the k-Fibonacci and k-Lucas Numbers *Int. J. Cont. Math. Scie.* 5(12) pp 569-578
[17] Zhang Y, Zhang H S and Chen G Y 2013 A note on the square roots of a class of circulant matrices *Jour. of Appl. Math.* 2013(601243) pp 1-6
[18] Ipek A 2011 On the spectral norms of circulant matrices with classical Fibonacci and Lucas numbers entries *Appl. Math. and Comp.* 217(12) pp 6011-6012

Acknowledgments

This work was supported by the National Natural Science Foundation of China under Grant Nos. 61402071 and 61671099, Liaoning Province Nature Science Foundation under Grant Nos. 2015020006 and 2015020011, and the Fundamental Research Funds for the Central Universities under Grant Nos. 3132015230 and 3132016111.

A new spectral method using legendre wavelets for shallow water model in limited-area

Fukang Yin, Junqiang Song, Jianping Wu and Xiaoqun Cao

Academy of Ocean and Engineering, National University of Defense Technology, Changsha, P.R. China, 410073

yinfukang@nudt.edu.cn

Abstract. This paper presents a new spectral method using Legendre wavelets (named LWSTCM), which complete the stepping in spectral space while deal with boundary conditions in grid-point space by collocation method, for the numerical solution of shallow water model in limited-area. In order to deal with the overlapping boundaries, some proper schemes are considered for exchanging the information on the boundaries between sub-domains. 1-D advection equation is used to analysis the exponential convergence property and error characteristics of LWSTCM. Finally, we study LWSTCM on 2-D shallow water equations for a more realistic application. The numerical results are compared with existing numerical solutions found in the literature and demonstrate the validity and applicability of the presented method.

1. Introduction

The most common numerical methods used for atmospheric and oceanic models are finite difference method (FDM), finite elements method (FEM) and spectral method. Moreover, FDM is an approximation to the differential equation while others are an approximation to its solution. Comparing with FDM and FEM, spectral method can eliminate pole problems and has high accuracy due to its "exponential-convergence" property [1]. The spectral methods also offer the discrete conservations of kinetic energy and enstropy, which are very important for the 2-D turbulence modeling [2].

As a relatively new and an emerging area in mathematical research, wavelet theory has been applied in many different fields of science and engineering. Wavelets permit the accurate representation of a variety of functions and operators. Moreover, wavelets establish a connection with fast numerical algorithms [3]. Therefore, various wavelets [4-13] have been used for studying problems with greater computational complexity and proved to be powerful tools to explore a new direction in solving differential equations. Recently, spectral method using Legendre wavelets is successfully used to obtain the numerical solution of IVP on large intervals [14] and Klein/Sine-Gordon equations [15]. The main attraction of this method is that it can exploit multi-level parallelism by employing the multi-scale analysis and hierarchy structure of Legendre wavelets.

The research of the spectral methods for limited-area model always attracts many attentions. Tatsumi [16] developed a spectral limited-area model by employing sinusoidal-subtracted Fourier sine-cosine series-expansion method. Hoyer [17] divided the basic variables into background field and perturbation components and deal with the perturbation part with spectral methods through double trigonometric series. Fulton, S. R. and Schubert, W. H. [18,19] proposed a Chebyshev spectral method for the limited-area atmospheric and oceanic modeling. Haugen and Machenhauer [20] used Fourier series with cyclic boundary conditions in high resolution limited-area model; then, harmonic-sine series and cosine series expansion methods for limited-area model were develop by Chen and Kuo in [21] and Chen and Bai in [22] respectively. Hereafter, Juang and Kanamitsu [23] developed NCEP regional spectral model and Chen and Bai (1996) [24] developed a harmonic-Fourier spectral limited-area model with an external wind lateral boundary condition. Kuo [25] discussed the scale-dependent accuracy associated with the regional spectral model variables expanded by sine-cosine series. Katsuyuki V.Ooyama [26] gave some tests in 1-D single domain using cubic-spline spectral transform method for the time integration of nonlinear meteorological equations.

The major issue in large scale problems arising from scientific and engineering computing is that of reducing the computational cost while preserving numerical accuracy. The purpose of this paper is to develop a new effectively and exponential convergent method, which combining the spectral tau method with collocation method using the Legendre wavelets as the basis, for the numerical solution of limited-area models. Section 2 introduces a new spectral method a using Legendre wavelets. A simple linear advection equation is adopted to explore the numerical accuracy of proposed method in section 3. In section 4, LWSTCM is applied to shallow water model. Conclusions and summary are made in section 5.

2. A new spectral method using legendre wavelets

Because of that Legendre wavelets basis doesn't satisfy the boundary conditions and the locality of Legendre wavelets, Legendre wavelets spectral method needs to exchange the information on the boundaries between sub-domains. The overset method is employed at those overlapped boundaries. Let k and M is the scale and order of Legendre wavelets, respectively. Let i denoted the ith sub-domains ($i=1,\ 2,\ \cdots,\ 2^{k-1}-1$). Five simple overset schemes are tested, including (1) $u(iM)=u(iM+1)=\left(u(iM+2)+u(iM-1)\right)/2$, (2) $u(iM+1)=u(iM)$ or $u(iM)=u(iM+1)$ (depending on the propagated direction of wave) (3) spline interpolation (one in each side) (4) spline interpolation (two in each side). The above four schemes is denoted by ib=1, ib=2, ib=3 and ib=4, respectively. Without loss of generality, the method which uses the exact solutions at the overlapped boundaries for comparing is noted as ib=5. The information exchange is show in figure 1 at the overlapped boundary, where the green points denote the points on overlapping boundaries, and the purple points denote the points that participate in the information exchange.

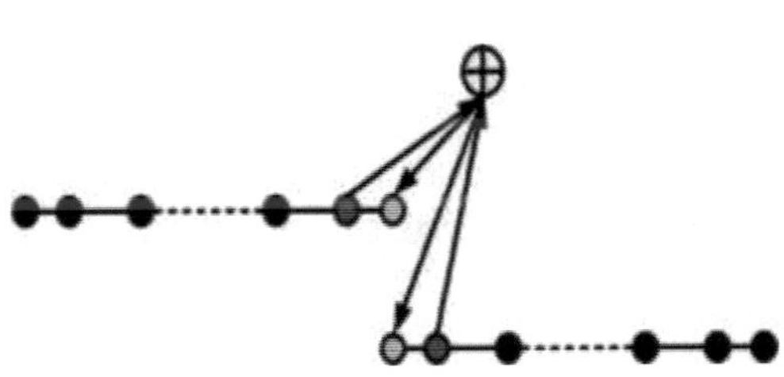

Figure 1. The information exchange at the overlapped boundary.

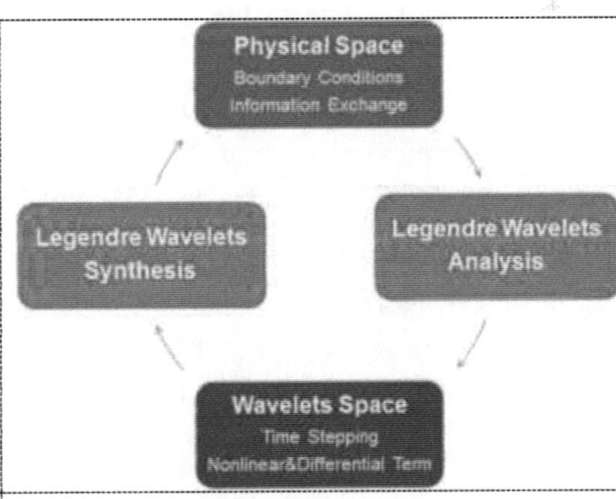

Figure 2. Schematic description of the spectral transform for LWSTCM.

As Fulton and Schubert point out, spectral collocation method is more stable than spectral Tau method. This is duo to the inaccuracy of treatment for the boundary conditions. What's more, it's nearly impossible to conduct the information exchange on the boundaries between sub-domains in spectral space. So we propose a hybrid method which complete time stepping in spectral space while deal with boundary conditions and information exchange in physical space by using collocation method. It should be noted that the collocation method is employed to evaluate the coefficient of Legendre wavelets expansion. Figure 2 gives the schematic description of the spectral transform for LWSTCM.

3. Test case

In this section, we give some computational results of one-dimensional linear advection equation to validation with the method based on preceding section, to support our theoretical discussion.
Consider the one-dimensional linear advection equation [1, 18, 25]

$$\frac{\partial u(x,t)}{\partial t}+\frac{\partial u(x,t)}{\partial x}=0, \tag{1}$$

in the domain $[-1,1]$. The exact solution of Eq. (1) is

$$u(x,t)=A\exp\left[-\left(\frac{x-x_0-t}{L}\right)^2\right]. \tag{2}$$

This is the simplest model involving wave or advective processes. The incoming boundary condition (1) is specified according to the exact solution at $x=-1$. No boundary condition is needed at right end point. This is an open boundary situation in the sense that any wave should propagated out of the domain without any deformation. In the case of that exact solution is Eq. (2), we assume $A=1.0$ and $x_0=-0.5$. In order to discuss the accuracy of LWSTCM, Chebyshev spectral collocation method (named CSCM) is introduced to do the comparison with it. It should be noted that the CSCM program used here is based on the burgers' code from the home page of Jie Shen (http://lsec.cc.ac.cn/~hyu/teaching/SHONM2013.html).
Figure 3 gives the L_2 errors (in log10 form) of single-scale LWSTCM as a function of the spatial-scale parameter L (Gaussian e-folding distance [1]) at $t=1.0$ for M =48. The single-scale LWSTCM clearly gives a much better approximation than CSCM method as L bigger than 0.2. In particular, the LWSTCM method reaches machine accuracy more slowly than CSCM. Figure 4 shows the corresponding L_2 errors of LWSTCM at $t=1.0$ as a function of M for different Legendre wavelets scale k and overlapped boundaries scheme. The exponential convergence of the two methods is obvious, while rapidly approach their asymptotic rate of decrease for L=2.0. From the results of multiple-scale LWSTCM in figure 4, it can be concluded that (1) the accuracy of overlapped boundaries scheme 2 is nearly the same as scheme 5 which using the exact solution, (2) the accuracy of overlapped boundaries scheme 1 is the same as scheme 3, (3) the overlapped boundaries scheme 2 is the best while the schemes 1 and 3 is the worst, (4) multiple-scale LWSTCM using scheme 2 is convergence fast with the increase of scale k.

Figure 5 presents the L_2 errors of single-scale LWSTCM as a function of M and dt at t =1.0. From the figure 5, it can be found that single-scale LWSTCM has good exponential convergence property and error characteristics.

Like the second test described in [18], we also consider the test case with following analytical solution

$$u(x,t)=\cos\left[\Omega\pi(x-t)\right], \tag{3}$$

where Ω is the wave number in the solution.
Figure 6 shows the L_2 error with dt= 0.0001 at $t=1.0$ as a function of M and Ω for the LWSTCM and CSCM. From figure 6, it can be found that LWSTCM requires less grid points than CSCM to achieve the same error for the same wave number while multiple-scale LWSTCM converges faster than single-scale LWSTCM.
The above results indicate that when the solution is smooth enough, LWSTCM needs less grid points than CSCM for the same accuracy and that this advantage increase dramatically as the desired accuracy increases. Even in the case of not smooth enough, LWSTCM has almost has the same accuracy as CSCM in the same degrees of freedom. Moreover, LWSTCM can save more storage and

computing time profit from the sparseness of the derivative matrix and the previous mentioned advantages.

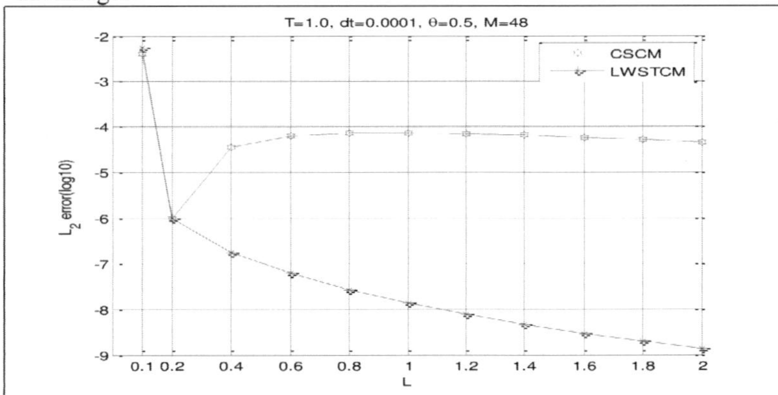

Figure 3. L_2 errors in the numerical solutions of the model problem 1 as a function of L at t=1.0 with dt=0.0001 for M=48 and k=0.

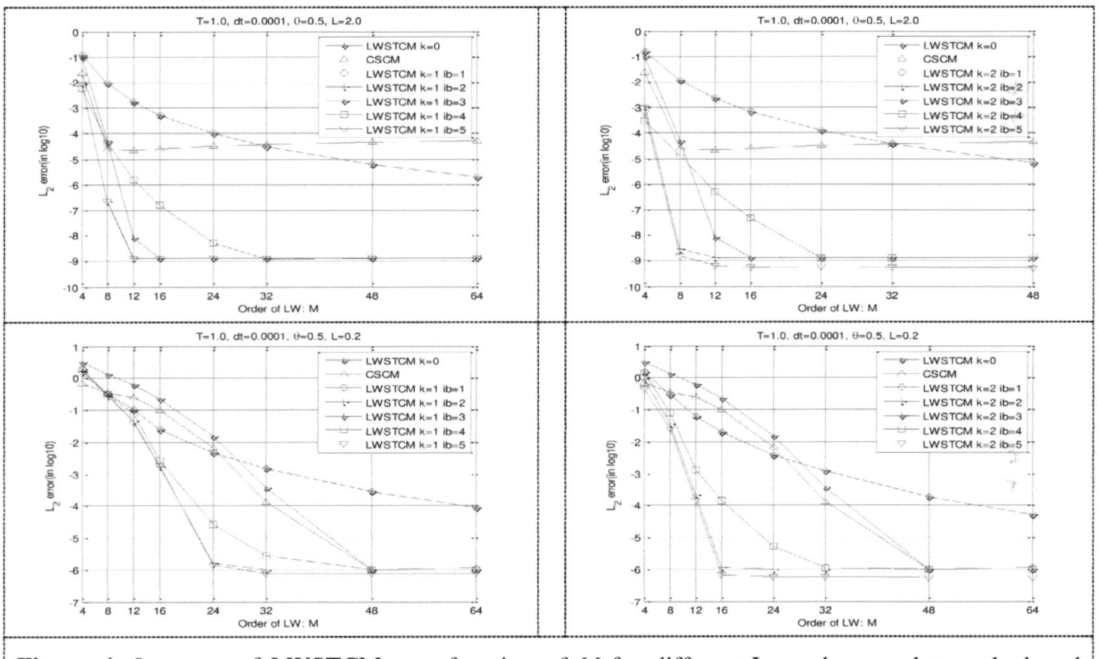

Figure 4. L_2 errors of LWSTCM as a function of M for different Legendre wavelets scale k and overlapped boundaries scheme at t =1.0 with dt=0.0001

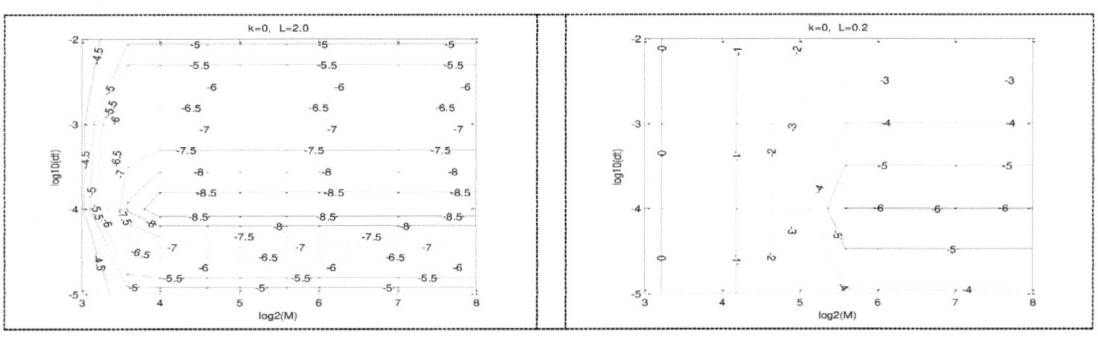

Figure 5. L_2 errors of single-scale LWSTCM as a function of M and dt at t =1.0 for L=2.0 and 0.2

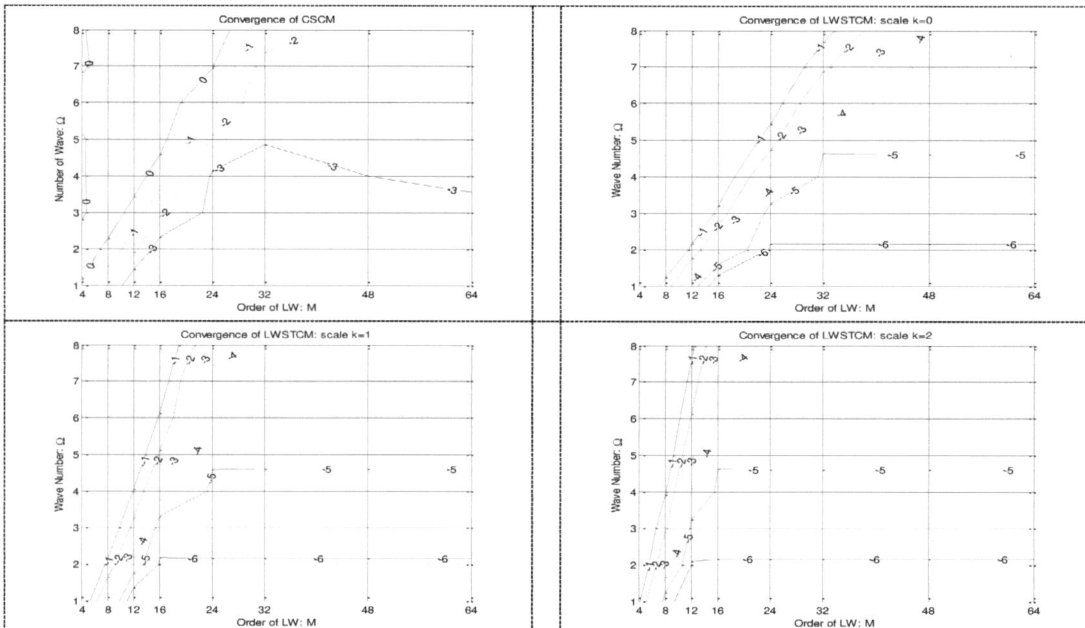

Figure 6. Logarithm of the L_2 errors in the numerical solutions of problem 1 for Eq. (5) as a function of M and Ω.

4. Limited-area shallow water model

In this section, the Houghton's shallow water model [27] is used to validate the feasibility and efficiency of the LWSTCM method. The domain is discretized using a mesh of $N_x \times N_y$ points, and fourth-order Adams-Bashforth scheme is used for time discretion. Let $\bar{u}$, $\bar{v}$ and $\bar{\phi}$ denotes the gird point value of function u, v and ϕ, respectively. The boundary conditions are set by a periodic overset condition along the east/west boundary and a wall-condition is applied at along the south and north boundaries. In other words,

$$\bar{u}(1,1:N_y)=\bar{u}(N_x-1,1:N_y),\ \bar{v}(1,1:N_y)=\bar{v}(N_x-1,1:N_y),\ \bar{\phi}(1,1:N_y)=\bar{\phi}(N_x-1,1:N_y), \tag{4}$$

while in the y-direction

$$\bar{v}(1:N_x,1)=\mathbf{O},\quad \bar{v}(1:N_x,N_y)=\mathbf{O}, \tag{5}$$

Note that no boundary conditions are necessary for u and ϕ in the y-direction. The integration time window was 24h with dt =10s. The time step can be increased by employing some technologies including more stable time discrete scheme and well-posed boundary conditions.

The initial conditions given by Grammeltvedt in reference [28] are

$$h(x,y)=H_0+H_1\tanh\left(9\frac{D/2-y}{2D}\right)+H_2\,\mathrm{sech}^2\left(9\frac{D/2-y}{2D}\right)\sin\left(\frac{2\pi x}{L}\right), \tag{6}$$

The initial velocity fields can be derived from the following relationship

$$u=-\frac{g}{f}\frac{\partial h}{\partial y},\quad v=\frac{g}{f}\frac{\partial h}{\partial x}. \tag{7}$$

The constants are listed as

$$L=4400\text{km},\ \ g=10\text{ms}^{-2},\ \ D=6000\text{km},\ \ H_2=133\text{m},$$
$$H_0=2000\text{m},\ \ \bar{f}=10^{-4}\text{s}^{-1},\ \ H_1=220\text{m},\ \ \beta\text{=}1.5\times10^{-11}\text{s}^{-1}\text{m}^{-1}, \tag{8}$$

Figures 7-9 show the geopotential field *h/c* at *t*=24, 48, and 72 hours for different (*M1,M2,k1,k2*). From figure 7, it can be found that the result of LWSTCM is the same as that obtained by using ADI finite-difference scheme [29]. By comparison with the results of single-scale LWSTCM, multiple-scale LWSTCM for Houghton's shallow water model shows some oscillations in geopotential field. These may be caused by the inaccuracy of the overlapping boundaries. From figures 7-9, it can be found that the results of multi-scale LWSTCM are good agree with those obtained by single scale LWSTCM when use the same number of grid points while have more degree of parallelism.

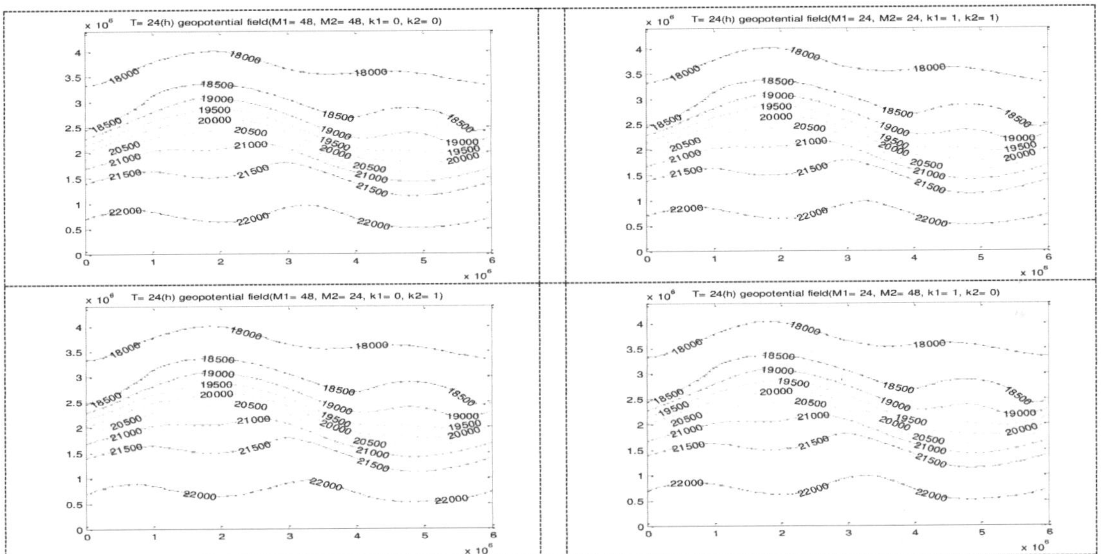

Figure 7. The geopotential field at *T*=24*h* for difference (*M1,M2,k1,k2*).

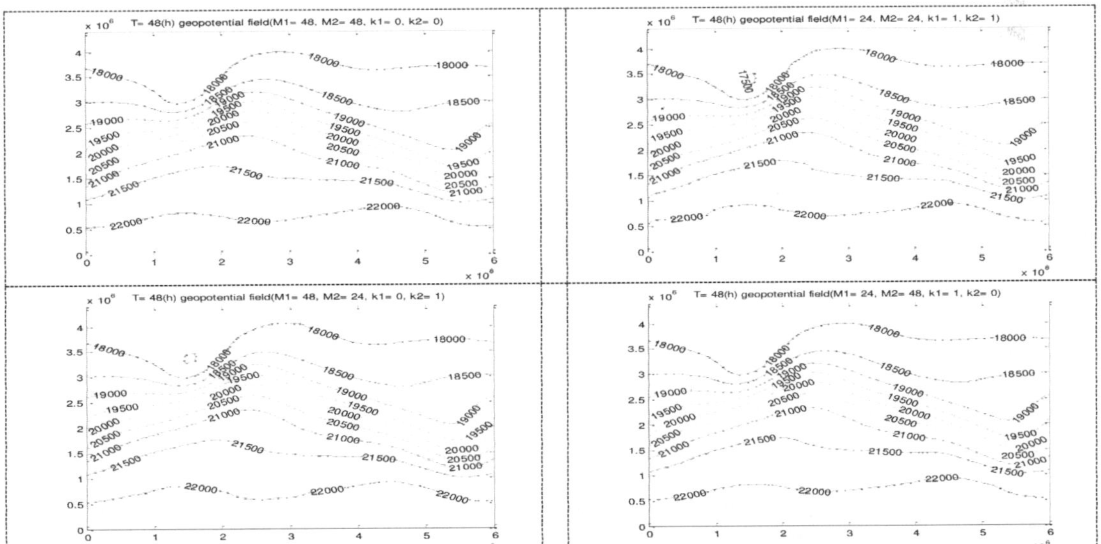

Figure 8. The geopotential field at *T*=48*h* for difference (*M1,M2,k1,k2*).

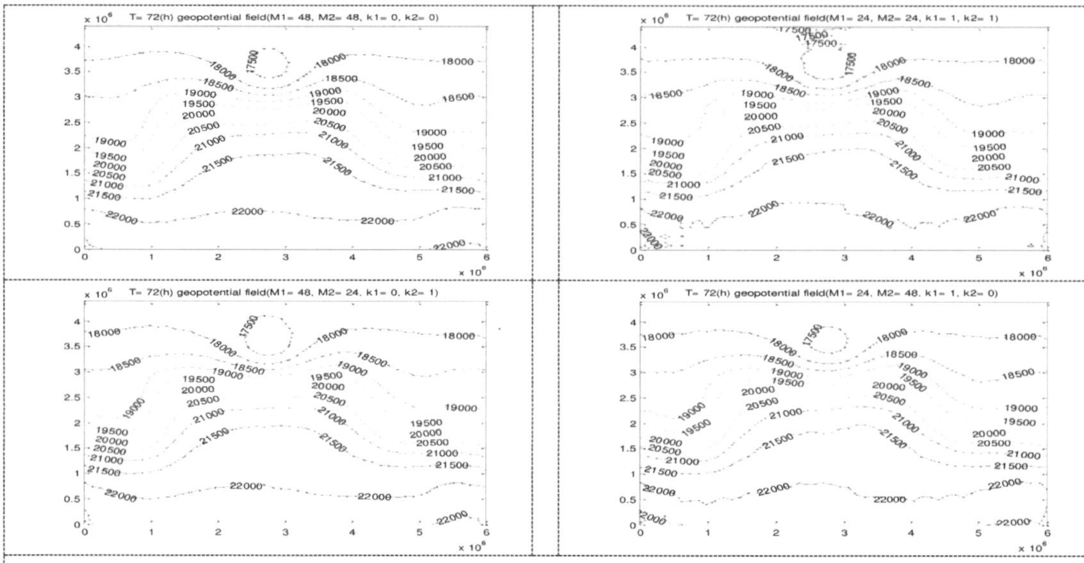

Figure 9. The geopotential field at T=72h for difference ($M1$,$M2$,$k1$,$k2$).

5. Conclusions

In this paper, we presented a hybrid spectral Tau method and collocation method for shallow water model in limited-area. The exponential convergence property and error characteristics are shown in the test of advection equation. LWSTCM clearly gives a much better approximation than CSCM as L (Gaussian e-folding distance) increases. In other words, LWSTCM needs less grid points than CSCM for the same accuracy and that this advantage increase dramatically as the desired accuracy increases. Further, LWSTCM appear to have better stability than CSCM for multiple-waves solutions. Finally, LWSTCM is applied to the solution of 2-D shallow water model for a more realistic application. Numerical results demonstrate that the LWSTCM can reduce the freedom of model and increase the parallelism of model while preserving numerical accuracy.

6. References

[1] Kuo, H. C., Williams, R. T., Boundary effects in regional spectral models. Mon. Wea. Rev., 120 (1992), 2986-2992.

[2] Yung-Chieh Chang, Chebyshev Collocation Method for Shallow Water Models with Domain Decomposition, PhD thesis, National Chiao Tung University, 2009.

[3] Beylkin, G., Coifman R., Rokhlin V., Fast wavelet transforms and numerical algorithms I. Commun. Pur. Appl. Math., 44 (1991), 141-183.

[4] Yousefi, S. A., Legendre wavelets method for solving differential equations of Lane-Emden type. Appl. Math. Comput., 181 (2006), 1417-1422.

[5] Lepik, Ü., Numerical solution of evolution equations by the Haar wavelet method. Appl. Math. Comput., 185 (2007), 695-704.

[6] Hariharan, G., Kannan K., Sharma K. R., Haar wavelet method for solving Fisher's equation. Appl. Math. Comput., 211 (2009), 284-292.

[7] Mohammadi, F., Hosseini M. M., A new Legendre wavelet operational matrix of derivative and its applications in solving the singular ordinary differential equations. J. Franklin I., 348 (2011), 1787-1796.

[8] Venkatesh, S. G., Ayyaswamy S. K., Raja Balachandar S., The Legendre wavelet method for solving initial value problems of Bratu-type. Comput. Math. Appl., 63 (2012), 1287-1295.

[9] Yang, C., and Hou J., Chebyshev wavelets method for solving Bratu's problem. Bound. Value. Probl., 2013 (2013), 142-146.

[10] Fukang Yin, Junqiang Song, and Fengshun Lu, A Coupled Method of Laplace Transform and Legendre Wavelets for Nonlinear Klein-Gordon Equations, Math. Method. Appl. Sci., 37 (2014) 781-792.

[11] Pandey R K, Kumar N, Bhardwaj A, et al. Solution of Lane-Emden type equations using legendre operational matrix of differentiation, Appl. Math. Comput., 218 (2012) 7629-7637.

[12] Jafari H, Yousefi S, Firoozjaee M, et al. Application of Legendre wavelets for solving fractional differential equations. Compu. Math. Appl., 62 (2011) 1038-1045.

[13] Liu N and Lin E-B, Legendre wavelet method for numerical solutions of partial differential equations, Numer. Meth. Part. D. E., 26 (2010) 81-94.

[14] A. Karimi Dizicheh, F. Ismail, M. Tavassoli Kajani, Mohammad Maleki, A Legendre wavelet spectral collocation method for solving oscillatory initial value problems, J. Appl. Math., 2013 (2013)1-5.

[15] Fukang Yin, Tian Tian, Junqiang Song, and Min Zhu, Spectral methods using Legendre wavelets for nonlinear Klein/Sine-Gordon equations, J. Comput. Appl. Math., 275 (2015) 321-334.

[16] Tatsumi, Y., A spectral limited-area model with time-dependent lateral boundary conditions and its application to a multi-level primitive equation model, J. Meteor. Soc. Japan, 64 (1986) 637-664.

[17] Hoyer, J. M., The ECMWF spectral limited-area model. Proc. ECMWF Workshop on Techniques for Horizontal Discretization in Numerical Weather Prediction Models, Berkshire, Reading, United Kingdom, ECMWF, 1987, 343-359.

[18] Fulton, S. R., and W. H. Schubert, Chebyshev spectral methods for limited-area models. Part I: Model problem analysis. Mon. Wea. Rev., 115 (1987) 1940-1953.

[19] Fulton, S. R., and W. H. Schubert, Chebyshev spectral methods for limited-area models. Part II: Shallow water model. Mon. Wea. Rev., 115 (1987) 1954-1965.

[20] Haugen, J. E., and Machenhauer, B., A spectral limited-area model formulation with time-dependent boundary conditions applied to the shallow-water equations. Mon. Wea. Rev., 121 (1993), 2618-2630.

[21] Chen Q. S., Kuo Y. H., A harmonic-sine series expansion and its application to partitioning and reconstruction problems in a limited area Mon. Wea. Rev., 120 (1992) 91-112.

[22] Chen Q. S., Kuo Y. H., A consistency condition for wind-field reconstruction in a limited area and a harmonic-cosine series expansion. Mon. Wea. Rev., 120 (1992) 2653-2670.

[23] Juang, H. M. H, and Kanamitsu M., The NMC nested regional spectral model. Mon. Wea. Rev., 122 (1994) 3-26.

[24] Chen, Q., Bai L., and Bromwich, D. H., A Harmonic-Fourier Spectral Limited-Area Model with an External Wind Lateral Boundary Condition. Mon. Wea. Rev., 125 (1997) 143-167.

[25] Kuo, H. C., and Williams, R. T., Scale-dependent accuracy in regional spectral methods. Mon. Wea. Rev., 126 (1998) 2640-2647.

[26] Ooyama K. V., The cubic-spline transform method: Basic definitions and tests in a 1D single domain. Mon. Wea. Rev., 130 (2002) 2392-2415.

[27] Houghton D, Kaaahara A, Washington W. Long-Term Integration of the Barotropic Equations by the Lax-Wendroff Method. Mon. Wea. Rev., 94 (1966) 141-150.

[28] Grammeltvedt A. A survey of Finite-Difference schemes for the primitive equations for a barotropic fluid. Mon. Wea. Rev., 97 (1969) 384-404.

[29] R. Ştefănescu and I.M. Navon, POD/DEIM nonlinear model order reduction of an ADI implicit shallow water equations model, J. Comput. Phys., 237 (2013) 95-114.

Acknowledgments
This work is supported by National Natural Science Foundation of China (Grant No. 61379022 and 41105063).

Finite time quenching for a nonlinear diffusion equation with singular boundary flux

Ying Yang

School of Mathematics and Statistics, Shenzhen University, Shenzhen, 518060, China

E-mail: yiyiying729@163.com

Abstract. In this paper, we study the finite time quenching phenomenon of a nonlinear diffusion equation with nonlinear source and singular boundary flux. It is shown that the solution must quench in a finite time and the time derivative blows up at a quenching point. The quenching set and rate are determined.

1. Introduction

This paper deals with the one dimensional nonlinear diffusion equation with nonlinear source and singular boundary flux

$$\begin{aligned}
&\frac{\partial A(u)}{\partial t} = u_{xx} + (1-u)^{-\alpha}, && 0<x<1, \quad t>0,\\
&u_x(0,t)=0, u_x(1,t) = -B(u(1,t)), && t>0, \qquad (1.1)\\
&u(x,0)=u_0(x), && 0 \le x \le 1.
\end{aligned}$$

Here $A(s)$ is an appropriately smooth and strictly monotone increasing function with $A(0)=0$, $A(1)=1$ and $A''(s) \le 0$ and $B(s)$ satisfies

$$B(s)>0, \quad B'(s)<0, \quad B''(s) \ge 0, \text{for} \quad s>0, \quad \lim_{s\to 0^+} B(s) = +\infty$$

The initial value $u_0(x):[0,1] \to (0,1)$ is nonincreasing and satisfies some compatibility conditions and α is a positive constant. For the case $A(u)=u^m$, the model (1.1) known as the classical porous medium equation which describes a number of physical phenomenon in the nature such as the flow of an isentropic gas through a porous medium ([1][2]) and heat transfer or diffusion.

The main purpose of this paper is to study the quenching phenomenon of solutions for the problem (1.1), namely, there exists a finite time T such that

$$\lim_{t\to T^-} \max\{u(x,t): 0 \le x \le 1\} \to 1 \quad \text{or} \quad \lim_{t\to T^-} \min\{u(x,t): 0 \le x \le 1\} \to 0.$$

From now on, we denote the quenching time of problem (1.1) with T.

The concept of quenching was first introduced by Kawarada [3] for a semilinear heat equation with reaction $u_t = u_{xx} + 1/(1-u)$ and the author proved that not only the reaction term, but also the time derivative blows up wherever u reaches level $u=1$. From then on, the quenching phenomena of solutions for evolutionary equations have been studied extensively. For example, in [4], Selcuk and Ozalp studied the semilinear heat equation with singular interior source and boundary condition. They prove that the solution quench in a finite time on the boundary and the time derivative blows up at the quenching point. The quenching rate and a lower bound for the quenching time are obtained. In [5], Chan and Treeyaprasert investigated the semilinear with a moving nonlinear source $\delta(x-vt)f(u(x,t))$ on a semi-infinite interval $(0,\infty)\times(0,T]$. They established the uniqueness of a nonnegative continuous solution $u(vt,t)$ which is a strictly increasing function of t and discussed that if u exists for $t\in[0,t_q)$ with $t_q<\infty$, then $\sup\{u(x,t):0\le x<\infty\}$ reaches c^- at t_q. For more results, see [6]–[10] and the references therein.

In recent years, more and more researchers focus on the quenching phenomenon for nonlinear diffusion equations, especially for singular and degenerate parabolic equations. Dao and Díaz [11] showed the global quenching phenomenon for L_1 initial datums of nonnegative solutions to the one dimensional degenerate parabolic problems $u_t-(|u_x|^{p-2}u_x)_x+\chi_{\{u>0\}}u^{-\beta}=0$, in which contained a singular absorption term. They proved the solutions occurs quenching in a finite time and obtained the upper and lower bounds at the quenching time. It is well known that the most striking manifestation of this nonlinear degeneracy is that in porous medium flow there is a finite speed of propagation of disturbances from rest which is more consistent with biological phenomena in the real world. It was the work by Nie, Wang and Zhou [12] concerned the quenching phenomenon of the solution to the porous medium equation. They shown the critical length for the quenching phenomenon occurs and got the location of the quenching points and the blowing up of the derivative of the solution with respect to the time at the quenching time. For more works of the quenching phenomenon for singular and degenerate parabolic equations, see [13]–[16] and the references therein.

In this paper, we focus our attention on the finite time quenching for one-dimensional nonlinear diffusion equation with nonlinear source and singular boundary flux. To the best of our knowledge, there is little work devoted to the quenching phenomena of degenerate parabolic equations with two nonlinear heat sources. Even for linear diffusion equations, the results are limited, see ([17]–[19]) etc. Quenching phenomenon is dependent on the singular term of the model and different singular terms cause the problem may occur quenching phenomenon at different levels. Obviously, in our problem (1.1), both the nonlinear source term and the outflux may become singular in some finite time and the presence of the nonlinear diffusive term in the equation make the discussion becomes more complicated. Motivated by the work [4], we will study the finite time quenching phenomenon of the solution and the time derivative blows up at a quenching point. Then, we will discuss the quenching rate and the lower bound for the quenching time.

Throughout this paper, we assume that

(H_1) $u_0''(x)+(1-u_0(x))^{-\alpha}\ge 0;$

(H_2) $u_0'(x)\le -xB(u_0(x)),\qquad 0\le x\le 1.$

The main results are stated as follows.

Theorem 1.1. Assume (H_1) holds. Then there exists a finite time T, such that the solution u of (1.1) quenches in this time, and $x=0$ is the unique quenching point.

Theorem 1.2. Assume that (H_1) holds and $\alpha\ge 1$. Then u_t blow up at the quenching point $x=0$.

Theorem 1.3. Assume that (H_1) and (H_2) hold. Then there exists a positive constant C_1 such that

$$u(0,t)\ge 1-C_1(T-t)^{1/(\alpha+1)}$$

as t sufficiently close to T.

This paper is arranged as follows. The quenching phenomenon in finite time of the solution for the problem (1.1) and the blow up of u_t at the quenching point $x=0$ will be shown in Section 2. Then, in Section 3, we establish estimates for the quenching rate and the lower bound for the quenching time.

2. Quenching on the boundary and blow-up of u_t

In this section, we prove finite time quenching for the solution. Due to the degeneracy of the equation, the considered solutions might not be classical in general and we should discuss weak solutions. However, for simplicity of our arguments, we assume that the solution is appropriately smooth, since we may consider some approximate boundary and initial value conditions.

Remark 2.1. We assume that the condition (H_1) is proper. Namely, we can easily construct such an initial function satisfying (H_1) and compatibility conditions. Let $u_0(x) = 0.9 - \frac{3}{2}x^{4.5}$. For example, for $\alpha = 9$ and $B(u) = u^{-\log_{30/7} 3}$, $u_0(x)$ satisfies (H_1) and compatibility conditions.

We then rewrite problem (1.1) into the following form:

$$\begin{aligned} &u_t = \gamma(u)u_{xx} + \gamma(u)(1-u)^{-\alpha}, && 0<x<1, \quad t>0, \\ &u_x(0,t) = 0, u_x(1,t) = -B(u(1,t)), && t>0, \\ &u(x,0) = u_0(x), && 0 \le x \le 1. \end{aligned} \tag{2.1}$$

Where $\gamma(u) = 1/A'(u)$.

In order to obtain the quenching phenomenon in finite time, we firstly need the following

Lemma 2.1. Assume that (H_1) hold and, for some $T_0 > 0$, the solution u of (2.1) exists in $(0,T_0)$. Then $u_x(x,t) < 0$ and $u_t(x,t) \ge 0$ in $(0,1] \times (0,T_0)$.

Proof. Denote $a = u_x$. Then we have

$$\begin{aligned} &a_t = \gamma(u)a_{xx} + \gamma'(u)u_{xx}a + \gamma'(u)(1-u)^{-\alpha}a + \alpha\gamma(u)(1-u)^{-\alpha-1}a, && 0<x<1, \quad 0<t<T_0, \\ &a(0,t) = 0, \quad a(1,t) = -B(u(1,t)), && 0<t<T_0, \\ &a(x,0) = u_0'(x), && 0 \le x \le 1. \end{aligned} \tag{2.2}$$

The maximum principle leads to $a < 0$, and thus $u_x(x,t) < 0$ in $(0,1] \times (0,T_0)$. So, in $(0,1] \times (0,T_0)$, we can see that the problem (2.2) is nondegenerate and $a = u_x$ is a classical solution for (2.2).
Similarly, letting $b = u_t$, then b meets

$$\begin{aligned} &b_t = \gamma(u)b_{xx} + \gamma'(u)u_{xx}b + \gamma'(u)(1-u)^{-\alpha}b + \alpha\gamma(u)(1-u)^{-\alpha-1}b, && 0<x<1, \quad 0<t<T_0, \\ &b_x(0,t) = 0, \quad b_x(1,t) = -B'(u(1,t))b(1,t), && 0<t<T_0, \\ &b(x,0) = \gamma(u_0(x))u_0''(x) + \gamma(u_0(x))(1-u_0(x))^{-\alpha}, && 0 \le x \le 1. \end{aligned}$$

Making use of the maximum principle, we also obtain $u_t(x,t) \ge 0$ in $(0,1] \times (0,T_0)$. Therefore, the solutions of the problem (1.1) $u \in C^{2,1}((0,1] \times (0,T_0))$ with $u_x(x,t) < 0$ and $u_t(x,t) \ge 0$ in $(0,1] \times (0,T_0)$.

In Theorem 1.1 followed, we research the quenching phenomenon of the problem (1.1).

Proof of Theorem 1.1. By the maximum principle, we know that $0 < u(\square,t) < 1$ for all t in $(0,T_0)$. Assume that $u_0(x)$ satisfies (H_1). Then we get

$$\delta = -B(u(1,0)) + \int_0^1 (1-u(x,0))^{-\alpha} dx > 0.$$

Define $D(t) = \int_0^1 (1 - A(u(x,t)))dx$. Then $D(t)$ satisfies

$$D'(t) = -\int_0^1 \frac{\partial A(t)}{\partial t} dx = B(u(1,t)) - \int_0^1 (1-u(x,t))^{-\alpha} dx \le -\delta,$$

by Lemma 2.1. So $D(t) \le D(0) - \delta t$, which implies that $D(t_0)$=0 for some $t_0 > 0$. From the fact that

$u_x(x,t)<0$ for $0<x\le 1$, we find that there exists $T(0<T<t_0)$ such that $\lim_{t\to T^-} u(0,t)=1$. By virtue of the singular nonlinearity in the source, u must quench at $x=0$. In what follows, we devote to prove that quenching cannot occur in $(0,1)\times(\zeta,T)$ for some $\zeta(0<\zeta<T)$.
Define

$$E(x,t)=u_x+\varepsilon(c_2-x), \qquad \text{in} \quad (c_1,c_2)\times[\zeta,T)$$

Where $c_2\in(0,1), c_1\in(0,c_2)$ and ε is a positive constant to be specified later. Then $E(x,t)$ satisfies

$$E_t=\gamma(u)E_{xx}+\frac{\gamma'(u)}{\gamma(u)}u_xu_t+\alpha\gamma(u)(1-u)^{-\alpha-1}u_x\le\gamma(u)E_{xx},$$

That is,

$$E_t-\gamma(u)E_{xx}\le 0, \qquad (c_1,c_2)\times[\zeta,T),$$

since $u_x(x,t)<0$ in $(0,1]\times[0,T)$. Thus, $E(x,t)$ cannot research a positive interior maximum by the maximum principle. Furthermore, if ε is small enough, $E(x,\zeta)<0$ since $u_x(x,t)<0$ in $(0,1]\times[0,T)$ and

$$E(c_1,t)=u_x(c_1,t)+\varepsilon(c_2-c_1)<0,$$
$$E(c_2,t)=u_x(c_2,t)<0,$$

for $t\in(\zeta,T)$. According to the maximum principle, we get $E(x,t)<0$, i.e., $u_x<-\varepsilon(c_2-x)$ for (x,t) $[c_1,c_2]\times[\zeta,T)$. Integrating this with respect to x from c_1 to c_2, we have

$$u(c_2,t)<u(c_1,t)-\frac{\varepsilon(c_2-c_1)^2}{2}<1-\frac{\varepsilon(c_2-c_1)^2}{2}<1,$$

which means u does not quench in $(0.1]$. The theorem is proved. □
Now, we are in a position to discuss the result that u_t blows up at the quenching point $x=0$.

Proof of Theorem 1.2. Suppose that u_t is bounded on $(0,1]\times[0,T)$. Then, there exists a positive constant C such that $u_t<C$. That is,

$$\gamma(u)(u_{xx}+(1-u)^{-\alpha})<C.$$

Since $A''(s)<0$, then we know $A'(s)$ is nonincreasing. By Lemma 2.1, we know $u\in C^{2,1}([0,1]\times[0,T))$, so there exists a ϖ and d_1 such that $0<d_1\le u<1$ in $[0,\varpi]\times[0,T)$. Therefore

$$\gamma(u)=\frac{1}{A'(u)}\ge\gamma(d_1).$$

Thus,

$$\gamma(d_1)(u_{xx}+(1-u)^{-\alpha})\le\gamma(u)(u_{xx}+(1-u)^{-\alpha})<C,$$

that is

$$u_{xx}+(1-u)^{-\alpha}\le\frac{C}{\gamma(d_1)}.$$

Multiplying this inequality by u_x and integrating with respect to x from 0 to x, ones have

$$\ln[1-u(0,t)]>-\frac{1}{2}u_x^2+\ln[1-u(x,t)]+\frac{C}{\gamma(d_1)}[u(x,t)-u(0,t)]$$

for $\alpha=1$ and

$$\frac{(1-u(0,t))^{-\alpha+1}}{-\alpha+1}>-\frac{1}{2}u_x^2+\frac{(1-u(x,t))^{-\alpha+1}}{-\alpha+1}+\frac{C}{\gamma(d_1)}[u(x,t)-u(0,t)]$$

for $\alpha\ne 1$. We obtain, as $t\to T^-$ and $\alpha\ge 1$, that the left-hand side tends to negative infinity, while the right-hand side is finite. This contradiction shows that u_t blow up at the quenching point $x=0$. □

3. Quenching rate and the lower bound of quenching time
In this section, we present bounds on the quenching rate and a lower bound for the quenching time.

Proof of Theorem 1.3. Denote $F(x,t)=u_x(x,t)+xB(u(x,t))$ in $(0,1]\times[0,T)$. Then, $F(x,t)$ satisfies

$$\begin{aligned} &F_t-\gamma(u)F_{xx} \\ &=\gamma'(u)u_x[u_{xx}+(1-u)^{-\alpha}]+\gamma(u)[\alpha(1-u)^{-\alpha-1}-2B'(u)]u_x \\ &\quad +xB'(u)\gamma(u)(1-u)^{-\alpha}-\gamma(u)xB''(u)u_x^2\le 0, \end{aligned}$$

with $u_x<0$, $F(x,t)$ cannot reach a positive interior maximum. On the parabolic boundary, by (H_2),

$$F(x,0)=u_x(x,0)+xB(u(x,0))\le 0,$$

and

$$F(0,t)=0, \qquad F(1,t)=0,$$

for $t\in(0,T)$. Making use of the maximum principle, we obtain that $F(x,t)\le 0$ for $(x,t)\in[0,1]\times[0,T)$. So

$$F_x(0,t)=\lim_{\theta\to 0^+}\frac{F(\theta,t)-F(0,t)}{\theta}=\lim_{\theta\to 0^+}\frac{F(\theta,t)}{\theta}\le 0.$$

Thus, we get

$$F_x(0,t)=u_{xx}(0,t)+B(u(0,t)).$$

Then,

$$\gamma(u(0,t))F_x(u(0,t))=u_t(0,t)-\gamma(u(0,t))(1-u(0,t))^{-\alpha}+\gamma(u(0,t))B(u(0,t))\le 0.$$

Therefore, we have

$$u_t(0,t)\le\gamma(u(0,t))(1-u(0,t))^{-\alpha}\le\frac{1}{A'(1)}(1-u(0,t))^{-\alpha}.$$

Integrating for t from t to T, we get

$$u(0,t)\ge 1-C_2(T-t)^{\frac{1}{\alpha+1}},$$

Where $C_2=(\frac{\alpha+1}{A'(1)})^{1/(\alpha+1)}$. This completes the proof of Theorem 1.3. □

Remark 3.1. We can calculate a lower bound for the quenching time. From Theorem 1.3, a lower bound is $A'(1)(1-u_0(0))^{\alpha+1}/(\alpha+1)$ for quenching time T. If we choose, as in Remark 2.1, $u_0(x)=0.9-\frac{2}{3}x^{4.5}$, then we have $T=10^{-11}$ for $\alpha=9$.

4. References

[1] L. S. Leibenzon 1930 The Motion of a Gas in a Porous Me-dium (*Russian Academy of Sciences, Moscow*)
[2] M. Muskat 1937 The Flow of Homogeneous Fluids through Porous Media (*McGraw-Hill, New York*)
[3] H. Kawarada 1975 On solutions of initial-boundary problem $u_t=u_{xx}+1/(1-u)$ *Publ. Res. Inst. Math. Kyoto Univ.* **10** 729–36
[4] B. Selcuk and N. Ozalp 2014 The quenching behavior of a semilinear heat equation with a singular boundary outflux *Quart. Appl. Math.* **72** 747–52
[5] C. Y. Chan and T. Treeyaprasert 2015 Existence, uniqueness and quenching for a parabolic problem with a moving nonlinear source on a semi-infinite interval *Dynam. Systems Appl.* **24** 135–42
[6] M. Fila and H. A. Levine 1993 Quenching on the boundary *Nonlinear Anal.* **21** 795–802
[7] H. A. Levine 1983 The quenching of solutions of linear parabolic and hyperbolic equations with nonlinear boundary conditions *SIAM J. Math. Anal.* **14** 1139–53
[8] H. A. Levine 1989 Quenching, nonquenching, and beyond quenching for solutions of some parabolic equations *Ann. Mat. Pura Appl.* **155** 243–60

[9] R. Ferreira, A. D. Pablo, F. Quir´os and J. D. Rossi 2006 Non-simultaneous quenching in a system of heat equations coupled on the boundary *Z. Angew. Math. Phys.* **57** 1–9
[10] C. H. Jin, J. X. Yin and X. P. Zhang 2009 Critical quenching exponents for heat equations coupled with nonlinear boundary flux *Commun. Math. Res.* **25** 88–96
[11] N. A. Dao and J. I. D´ıaz 2016 A gradient estimate to a degenerate parabolic equation with a singular absorption term: The global quenching phenomena *J. Math. Anal. Appl.* **437** 445–73
[12] Y. Y. Nie, C. P.Wang and Q. Zhou 2013 Quenching for singular and degenerate quasilinear diffusion equations *Electron. J. Differential Equations* **131** 14 pp
[13] K. Deng and M. Xu 1999 Quenching for a nonlinear diffusion equation with a singular boundary condition *Z. Angew. Math. Phys.* **50** 574–84
[14] Y. Yang, J. X. Yin and C. H. Jin 2010 A quenching phenomenon for one-dimensional p-Laplacian with singular boundary flux *Appl. Math. Lett.* **23** 955–59
[15] Y. Yang, J. X. Yin and C. H. Jin 2016 Quenching phenomenon of positive radial solutions for p-Laplacian with singular boundary flux *J. Dyn. Control Syst.* **22** 653–60
[16] Y. Yang 2015 Quenching phenomenon for a non-Newtonian filtration equation with singular boundary flux *Bound. Value Probl.* **2015: 233** 11 pp
[17] C. Y. Chan and S. I. Yuen 2001 Parabolic problems with nonlinear absorptions and releases at the boundaries *Appl. Math. Comput.* **121** 203–09
[18] Y. H. Zhi and C. L. Mu 2007 The quenching behavior of a nonlinear parabolic equation with nonlinear boundary outflux *Appl. Math. Comput.* **184** 624–30
[19] C. L. Zhao 2000 Blow-up and quenching for solutions of some parabolic equations (*Ph.D. thesis, Univ. of Louisiana, Lafayette*)

Acknowledgments
This work is partially supported by the National Science Foundation of China (11571240,11301345, 11501378), Guangdong Natural Science Foundation (2014A030310074) and Natural Science Foundation of SZU (201545). The author would like to thank the referees for revising the manuscript and for many valuable comments, which helped to clarify this paper.

Rational Chebyshev pseudospectral method for long-short wave equations

Zeting Liu and Shujuan Lv

School of Mathematics and Systems Science & LMIB, Beihang University, Beijing, China, 100191

E-mail: lzt_well@163.com, lsj@buaa.edu.cn

Abstract. We consider the initial boundary value problem of the Long-Short wave equations on the whole line. Firstly, a three level linear fully discrete pseudospectral scheme are structured based on central difference in time and rational Chebyshev functions in space which are orthogonal in the L^2 space with weight 1. Secondly, the first-order differential matrix about rational Chebyshev functions is derived by the first-order differential matrix of Chebyshev polynomials, the entries of the matrix are just Chebyshev polynomials and Chebyshev Gauss collocation points. Thirdly, the numerical implementations are described and numerical results for the rational Chebyshev pseudospectral scheme are verified that a second-order accuracy in time and spectral accuracy in space.

1. Introduction

Many problems in science and engineering arise in unbounded domains and scientists have paid much attention to spectral method due to its high accuracy over the last three decades. Spectral method for solving PDEs on unbounded domains can be essentially classified into four approaches: (i) Domain truncation: truncate unbounded domains to bounded domains and solve the PDEs on bounded domains supplemented with artificial or transparent boundary conditions, see [1,2]; (ii) Approximation by other, non-classical orthogonal systems [3], or by rational orthogonal systems, for example, image of classical Jacobi polynomials through a suitable mapping, we refer to [4-7]; (iii) Mapping: map unbounded domains to bounded domains and use standard spectral methods to solve the mapped PDEs in the bounded domains, see [8,9]; (iv) Approximation by classical orthogonal systems on unbounded domains, such as Laguerre or Hermite polynomials/functions, see [10-12].

For problems with exponentially decaying solutions, Hermite or Laguerre spectral method is the first choice, while for problems with algebraically decaying solutions, we should choose rational spectral method. As for rational spectral method, the rational functions are usually defined as the composite functions of Jacobi polynomials and the associated mappings, so they are orthogonal in the weighted space $L^2_\omega(\mathbf{R})$ with a non-uniform weight function ω. Though the weight is much weaker than the Hermite and Laguerre spectral methods, however, the non-uniform weights in the standard rational approximations may bring in some difficulties in actual computation and destroy equations' conservation laws which play important roles in theoretical analysis and numerical simulation, such as the Schödinger equation, the Korteweg-de Vries equation and so on. There are only a few papers using rational functions which are orthogonal in the usual space $L^2(\mathbf{R})$, for papers using Chebyshev rational functions we refer to [13,14], Legendre rational functions [15,16].

In this paper, we consider Chebyshev rational functions which absorb the weight in its definition and are orthogonal in the usual (non-weighted) space $L^2(R)$ for the following Long-Short wave (LS) equations which the solutions are algebraically decaying at infinity:

$$\begin{cases} is_t + s_{xx} = \alpha sl + f, & x \in \boldsymbol{R},\ 0 < t \le T, & (1.1) \\ l_t + \beta(|s|^2)_x = g, & x \in \boldsymbol{R},\ 0 < t \le T, & (1.2) \\ s(x,0) = s_0(x), l(x,0) = l_0(x), & x \in \boldsymbol{R}, & (1.3) \\ \lim\limits_{|x|\to\infty} s(x,t) = \lim\limits_{|x|\to\infty} s_x(x,t) = \lim\limits_{|x|\to\infty} l(x,t) = 0, 0 < t \le T. & & (1.4) \end{cases}$$

where complex function s is the envelope of the short wave, and real function l is the amplitude of the long wave and α,β are positive numbers, f and g are source terms(force terms). The LS equations are a kind of important nonlinear evolution equations in physics, which are found by Djordjevic and Redekopp[17]. They describe that the dispersion of the short wave is balanced by nonlinear interaction of the long wave, while the evolution of the long wave is driven by the self-interaction of the short wave. The LS equations appear in various physical situations, such as in the analysis of electron-plasma and ion-field interaction[18], plasma physics [19].
An outline of this paper is as follows. We introduce some notations and recall some properties about Chebyshev rational functions and interpolation, then we build a three linear fully discrete pseudospectral scheme in section 2. In section 3, we derive the first-order differential matrix about rational Chebyshev functions by the first-order differential matrix of Chebyshev polynomials. In section 4, we describe the numerical implementations and show the numerical results that a second-order accuracy in time and spectral accuracy in space.

2. Preliminaries and notations

Let $T_l(y)$ be the Chebyshev polynomial of degree l and it satisfies

$$\int_{-1}^{1} T_l(y)T_{m(y)}\omega(y)dy = c_l\delta_{lm} \tag{2.1}$$

where $\omega(y)=(1-y^2)^{-1/2}$, δ_{lm} is the Kronecker fuction, and $c_0 = \pi, c_l = \frac{\pi}{2}\ for\ l \ge 1$.
The Chebyshev rational functions of degree l is defined by

$$R_l(x) = \frac{1}{\sqrt{c_l}}\frac{1}{\sqrt{x^2+1}}T_l\left(\frac{x}{\sqrt{x^2+1}}\right), l = 0,1,2,\cdots.$$

The $R_l(x)$ is the eigenfunction of the singular Sturm-Liouville problem

$$(x^2+1)^{\frac{1}{2}}\partial_x\left((x^2+1)\partial_x\left((x^2+1)^{\frac{1}{2}}R_l(x)\right)\right) + l^2R_l(x) = 0, l = 0,1,2,\cdots \tag{2.2}$$

Using (2.1), we have

$$\int_{\boldsymbol{R}} R_l(x)R_m(x)dx = \delta_{lm} \tag{2.3}$$

For any $v \in L^2(\mathbf{R})$, we can write

$$v = \sum_{l=0}^{\infty} \hat{v}_l\, R_l(x), \hat{v}_l = \int_{\boldsymbol{R}} vR_l(x)dx \tag{2.4}$$

For any given positive integer N, let

$$T_N = span\{T_0(x), T_1(x), \cdots T_N(x)\}$$

and

$$\mathcal{R}_N = span\{R_0(x), R_1(x), \cdots, R_N(x)\}$$

In actual computations, we often use Rational Chebyshev-Gauss interpolation. We set $x_{N,j}, 0 \le j \le N$ be the collocation nodes which are the $N+1$ distinct zero points of R_{N+1}. Due to the relation of R_{N+1} and T_{N+1}, we obtain

$$x_{N,j} = -\cot\frac{\pi(2j+1)}{2N+2}, 0 \le j \le N$$

The corresponding weights are

$$\omega_{N,j} = \frac{\pi}{N+1}(x_{N,j}^2 + 1), 0 \le j \le N$$

The discrete inner product and norm are defined as follows,

$$(u,v) = \sum_{j=0}^{N} u(x_{N,j})v(x_{N,j})\omega x_{N,j}, \qquad \|v\|_N = (v,v)_N^{\frac{1}{2}}$$

It can be computed that for any $\Phi \in \mathcal{R}_i$, $\psi \in \mathcal{R}_j$ with $i + j \le 2N + 1$, we have

$$(\Phi,\psi) = (\Phi,\psi)_N$$

Let τ be the step-size in variable $t, t_k = k\tau(k = 0,1,\cdots,M; M = [T/\tau]), u^k = u(x,t_k)$, and

$$u_{\hat{t}}^k = \frac{u^{k+1} - u^{k-1}}{\tau}, \bar{u^k} = \frac{u^{k+1} + u^{k-1}}{2}$$

The fully discrete Chebyshev rational functions pseudospectral scheme for $(1.1) - (1.4)$ is to find $s_N^k(x,t), l_N^k(x,t) \in \mathcal{R}_N$ such that for any $v \in \mathcal{R}_N$, we have

$$\begin{cases} i\left(s_{N\hat{t}}^k, v\right) - \left(s_{Nx}^{\bar{k}}, v_x\right)_N = \alpha\left(s_N^{\bar{k}} l_N^k, v\right)_N + \left(f^{\bar{k}}, v\right)_N & (2.5) \\ \left(l_{N\hat{t}}^k, v\right) + \beta\left(\left|s_N^k\right|_x^2, v\right)_N = (g^k, v)_N & (2.6) \\ s_N^0 = I_N s_0, l_N^0 = I_N l_0 & (2.7) \\ s_N^1 = I_N(s_0 + i\tau(s_{0xx} - \alpha s_0 l_0 - f^0), l_N^1 = I_N\left(l_0 - \tau(\beta|s_0|_x^2 - g^0)\right) & (2.8) \end{cases}$$

3. The first-order differential matrix for rational Chebyshev functions

In this section, we are mainly derive the first-order differential matrix about rational Chebyshev functions by the first-order differential matrix of Chebyshev polynomials, the entries of the first-order differential matrix of Chebyshev polynomials are [20]:

$$d_{kj} = \begin{cases} \dfrac{T'_{N+1}(y_k)}{T'_{N+1}(y_j)}\dfrac{1}{y_k - y_j}, & if\ k \ne j \\ \dfrac{y_k}{2(1 - y_k^2)}, & if\ k = j \end{cases}$$

where $y_j(0 \le j \le N)$ are *N+1* distinct zeros of Chebyshev polynomial T_{N+1}.

Now we consider the first-order differential matrix about rational Chebyshev functions, it is obvious that for any $v \in \mathcal{R}_N$, we find $v = \frac{1}{\sqrt{x^2+1}}u(\frac{x}{\sqrt{x^2+1}})$ with $u \in \mathcal{T}_N$, Then we can compute its first-order derivative values by

$$\begin{aligned} v'(x_k) &= -\frac{x_k}{(x_k^2+1)^{\frac{3}{2}}}u\left(\frac{x}{\sqrt{x^2+1}}\right) + \frac{1}{(x_k^2+1)^2}u' \\ &= -\frac{x_k}{x_k^2+1}v + \frac{1}{(x_k^2+1)^2}\sum_{j=0}^{N} u\left(\frac{x_j}{\sqrt{x_j^2+1}}\right)d_{kj} \\ &= -\frac{x_k}{x_k^2+1}v + \frac{1}{(x_k^2+1)^2}\sum_{j=0}^{N}(x_j^2+1)^{\frac{1}{2}}v(x_j)d_{kj} \\ &:= \sum_{j=0}^{N} v(x_j)\hat{d}_{kj} \end{aligned}$$

where

$$\hat{d}_{kj} = \begin{cases} \frac{(x_j^2+1)^{\frac{1}{2}}}{(x_k^2+1)^2}\frac{T'_{N+1}(y_k)}{T'_{N+1}(y_j)}\frac{1}{y_k-y_j}, & if\ k \neq j, \\ -\frac{x_k}{x_k^2+1}+\frac{1}{(x_k^2+1)^{\frac{3}{2}}}\frac{y_k}{2(1-y_k^2)}, & if\ k = j. \end{cases}$$

After uniforming variables, we obtain

$$\hat{d}_{kj} = \begin{cases} \frac{(1-y_j^2)^2}{(1-y_j^2)^{\frac{1}{2}}}\frac{T'_{N+1}(y_k)}{T'_{N+1}(y_j)}\frac{1}{y_k-y_j}, & if\ k \neq j, \\ -\frac{1}{2}y_k(1-y_k^2)^{\frac{1}{2}}, & if\ k = j. \end{cases}$$

Thus we get the first-order differential matrix about rational Chebyshev functions, the entries of the matrix are just about Chebyshev polynomials and Chebyshev Gauss collocation points.

4. Numerical experiments

We consider LS equations $(1.1)-(1.4)$ with $\alpha = \beta = 1$ and the following source terms:
$f(x,t) = (-\frac{6+12ix}{(x^2+1)^4}+\frac{48x^2}{(x^2+1)^5}-\frac{\sin t}{(x^2+1)^6})e^{i(x-t)}$ and $g(x,t) = \frac{\cos t}{(x^2+1)^3}-\frac{12x}{(x^2+1)^7}$

The exact solutions of this example are:

$$s(x,t) = \frac{e^{i(x-t)}}{(x^2+1)^3}\ \ and\ \ l(x,t) = \frac{\sin t}{(x^2+1)^3}$$

For the rational Chebyshev functions pseudospectral method, we choose Lagrange functions with weights $\omega_m(x) = \left(\frac{x_m^2+1}{x^2+1}\right)^{1/2}$ and denote $h_m(x) = \ell_m\omega_m(x)$, then we rewrite the numerical solutions as

$$s_N^{k+1}(x) = \sum_{m=0}^{N} s_m^{k+1}h_m(x), \quad l_N^{k+1}(x) = \sum_{n=0}^{N} l_n^{k+1}(x)$$

where $s_m^{k+1} = s_N^{k+1}(x_m)\ and\ l_n^{k+1} = l_N^{k+1}(x_n)$ are the nodal values of discrete solutions.
Then we obtain the following system of linear algebraic equations:

$$\left(iA-\frac{\tau}{2}B-\frac{\tau}{2}C\right)s^{k+1} = \left(iA+\frac{\tau}{2}B+\frac{\tau}{2}C\right)s^k+\tau A\boldsymbol{f}^{k+\frac{1}{2}}$$

$$\boldsymbol{l}^{k+1} = \boldsymbol{l}^k - 2\tau\mathrm{Re}Es^k + \tau\boldsymbol{g}^{k+\frac{1}{2}}$$

where

$$\boldsymbol{s}^{k+1} = \left(s_0^{k+1}, s_1^{k+1}, \cdots, s_n^{k+1}\right)^T, \qquad \boldsymbol{l}^{k+1} = \left(l_0^{k+1}, l_1^{k+1}, \cdots, l_N^{k+1}\right)^T$$

$$\boldsymbol{f}^k = (f^{k+\frac{1}{2}}(x_0), f^{k+\frac{1}{2}}(x_1), \cdots, f^{k+\frac{1}{2}}(x_N))^T, \qquad \boldsymbol{g}^k = (g^{k+\frac{1}{2}}(x_0), g^{k+\frac{1}{2}}(x_1), \cdots, g^{k+\frac{1}{2}}(x_N))^T$$

$$A = (a_{ij})_{i,j} = 0,1,\cdots,N, \qquad a_{ij} = \omega_j\delta_{ij}, \qquad B = D^TAD$$

$D = (\hat{d}_{ij})$ is the first-order differential matrix of rational Chebyshev functions,

$$C = diag(l_0^k\omega_0, l_1^k w_1, \cdots, l_N^k\omega_N), \qquad E = diag(\overline{s^k})D$$

In order to see the convergence order for the pseudospectral scheme, we present the numerical results below

Table 1. Errors for short wave s and long wave l at *N=128*.

τ	$s-s_N$		$l-l_N$	
	$L^2-error$	$L^\infty-error$	$L^2-error$	$L^\infty-error$
10^{-1}	2.5439e-02	3.2657e-03	3.6157e-02	8.5619e-03
10^{-2}	2.5735e-04	3.3063e-05	4.2118e-04	9,9175e-05
10^{-3}	3.0997e-06	5.9260e-07	4.3349e-06	1.1312e-06

From Table 1, we clearly see that both $L^2-error\ and\ L^\infty-error$ are at least satisfied second-order accuracy in time at different τ for given *N*.

Table 2. Errors for short wave s and long wave l at $\tau = 0.0001$.

τ	$s - s_N$		$l - l_N$	
	$L^2 - error$	$L^\infty - error$	$L^2 - error$	$L^\infty - error$
8	1.5007e-01	8,7101e-02	1.0811e-01	7.4045e-02
16	7.7535e-04	4.3104e-04	2.4714e-04	5.7528e-04
32	8.6726e-05	3.0869e-05	6.9407e-05	2.7524e-05
64	1.8885e-05	2.9068e-06	5.8743e-06	1.8534e-06
128	1.7463e-06	3.9237e-07	8.0942e-07	1.6435e-07

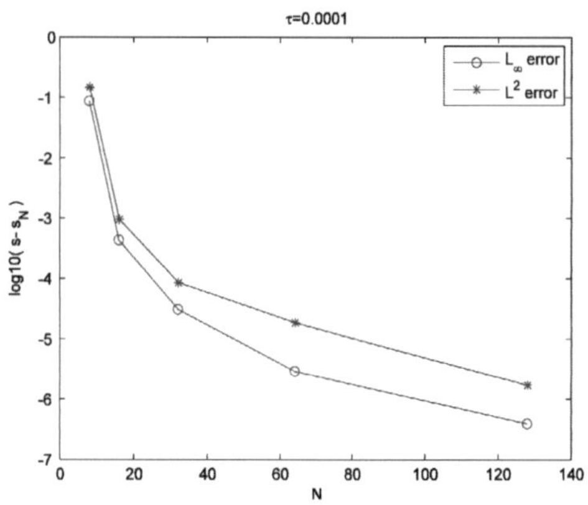

Figure 1. The errors for short wave.

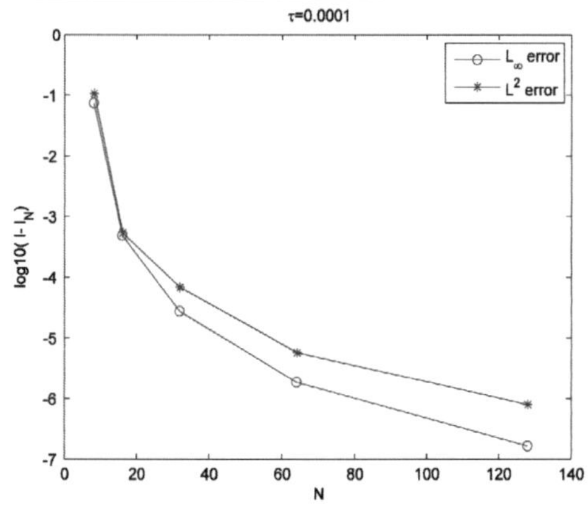

Figure 2. The errors for long wave.

From Table 2 and figure 1 and figure 2, we find that both $L^2 - error$ and $L^\infty - error$ are satisfied spectral accuracy in space at different N for given τ.

5. Conclusion

We use rational Chebyshev functions which are orthogonal in the usual L^2 space as base functions to solve LS equations on the whole line. We construct a three level linear explicit scheme, derive the first-order differential matrix about rational Chebyshev functions, whose entries of the matrix are just about Chebyshev polynomials and Chebyshev Gauss collocation points, and finally we present numerical results which confirm the scheme we constructed is second-order in time and spectral accuracy in space. The numerical convergence rates indicate that pseudospectral method using rational Chebyshev functions are very effective tools for numerical solutions of PDEs on the whole line.

6. Reference

[1] Shen J and Wang L L 2007 Analysis of a spectral-Galerkin approximation to the Helmholtz equation in exterior domains *SIAM J. Numer. Anal.* **45** pp 1954-78

[2] Nicholls D and Shen J 2006 A stable high-order method for two-dimensional bounded-obstacal scattering *SIAM J. Sci. Comput.* **28** pp 1398-1419

[3] Christov C I 1982 A complete orthonormal system of functions in $L^2(-\infty,+\infty)$ space SIAM J. Appl. Math. **42** pp 1337-44

[4] Guo B Y, Shen J and Wang Z Q 2000 A rational approximation and its applications to differential equations on the half line *J. Sci. Comp.* **15** 117-47

[5] Zhang Z Q and Ma H P 2009 A rational spectral method for the KdV equation on the half line *J. Comput. Appl. Math.* **230** pp 614-25

[6] Guo B Y and Yi Y G 2010 Generalized Jacobi rational spectral method and its applications, *J. Sci. Comput.* **43** pp 201-38

[7] Yi Y G and Guo B Y 2012 Generalized Jacobi rational spectral method on the half line *Adv. Comput. Math.* **37** pp 1-37

[8] Boyd J P 1987 Orthogonal rational functions on a semi-infinite interval *J. Comput. Phys.* **70** pp 63-88

[9] Boyd J P 1987 Spectral methods using rational basis functions on an infinite interval *J. Comput. Phys.* **69** pp 112-42

[10] Grosch C E and Orszag S A 1977 Numerical solution of problems in unbounded regions:coordinates transforms, *J. Comput. Phys.* **25** pp 273-96

[11] Shen J 2000 Stable and efficient spectral methods in unbounded domains using Laguerre functions *SIAM J. Numer. Anal.* **38** pp 1113-33

[12] Ma H P, Sun W W and Tang T 2005 Hermite spectral methods with time-dependent scaling for parabolic equations in unbounded domains *SIAM J. Numer. Anal.* **43** pp 58-75

[13] Guo B Y and Wang Z Q 2003 Modified Chebyshev rational spectral method for the whole line Discrete *Contin. Dyn. Syst.* **suppl** pp 365-74

[14] Shen J, Wang L L and Yu H J 2014 Approximations by orthonormal mapped Chebyshev functions for higher-dimensional problems in unbounded domains *J. Comput. Appl. Math.* **265** pp 264-75

[15] Guo B Y and Shen J 2001 On Spectral approximations using modified Legendre rational functions:application to the Korteweg-De Vries equation on the half line, *Indiana U. Math. J.* **50** pp 181-204

[16] Wang Z Q and Guo B Y 2004 Modified Legendre rational spectral method for the whole line *J. Comput. Math.* **22** pp 457-74

[17] Djordjevic V D and Redekop L G 1977 On two-dimensional packets of capillary-gravity waves *J. Fluid Mech.* **79** pp 703-14

[18] Nishikawa K, Mima H H and Ikezi K H 1974 Coupled nonlinear electron-plasma and ion acoustic waves *Phys. Rev. Lett.* **33** pp 148-51

[19] Nicholson D R and Goldman M V 1976 Damped nonlinear Schrödinger equation *Phys. Fluid* **19** pp 1621-25

[20] Shen J, Tang T and Wang L L 2011 *Spectral methods: Algorithms, Analysis and Applications* (Berlin Heidelberg: Springer-Verlag)

Acknowledgements
The authors were supported by the NSF of China (No. 11272024 and No. 11672011).

IOP Conf. Series: Journal of Physics: Conf. Series **814** (2017) 012017 doi:10.1088/1742-6596/814/1/012017

An accelerated randomized extended Kaczmarz algorithm

Xu Xiang[1,*], Xu Liu[1], Wentang Tan[1] and Xiang Dai[1]

[1]Southwest Electronics and Telecommunication Technology Research Institute, Chengdu, Sichuan, China, 610041

[*] E-mail address: xiangxu1402@163.com

Abstract. The randomized Kaczmarz(RK) is a useful algorithm for solving consistent linear system $A\boldsymbol{x}=\boldsymbol{b}(A\in\mathbb{R}^{m\times n},\ \boldsymbol{b}\in\mathbb{R})$. It was proved that for inconsistent linear system, with randomized orthogonal projection, the randomized extended Kaczmarz(REK) method converges with an expected exponential rate. We describe an accelerated randomized extended Kaczmarz algorithm(AREK) with Nesterov's accelerated procedure. The analysis shows that AREK converges better than REK when A is dense and the smallest singular value of $A^T A$ is small.

1. Introduction

The Kaczmarz method[1] is a popular algorithm for solving overdetermined linear systems and has numerous applications from tomography to image processing[2, 3]. This method sweeps through the rows of A in a cyclic manner, projecting in each substep the last iterate orthogonally onto the solution hyperplane of $\langle \boldsymbol{a}_i, \boldsymbol{x}\rangle = \boldsymbol{b}_i$; then the $(k+1)^{th}$ estimate $\boldsymbol{x}_{k+1}$ is defined by

$$\boldsymbol{x}_{k+1} = \boldsymbol{x}_k + \frac{\boldsymbol{b}_i - \langle \boldsymbol{a}_i, \boldsymbol{x}_k\rangle}{\|\boldsymbol{a}_i\|_2^2}\boldsymbol{a}_i, \tag{1}$$

where i = k mod m + 1, $\boldsymbol{a}_i$ is the ith row of A.

T. Strohmer and R. Vershynin [4, 5] propose at each iteration to randomly select a row of A with probability proportional to Euclidean norm of the row. This method is the randomized Kaczmarz(RK) which can thus be described by

$$\boldsymbol{x}_{k+1} = \boldsymbol{x}_k + \frac{\boldsymbol{b}[p(i)] - \langle \boldsymbol{a}_{p(i)}, \boldsymbol{x}_k\rangle}{\|\boldsymbol{a}_{p(i)}\|_2^2}\boldsymbol{a}_{p(i)}, \tag{2}$$

where $p(i)$ takes values in $\{1,...,m\}$ with probabilities $\|\boldsymbol{a}_{p(i)}\|_2^2 / \|A\|_F^2$. This randomized version of the Kaczmarz method provides clear advantages over the standard method in many cases. And in [5], using the selection above,they were able to provide a proof for the expected rate of convergence,

$$E\|\boldsymbol{x}_k - \boldsymbol{x}^*\|_2^2 \le \left(1 - \frac{1}{R}\right)^k \cdot \|\boldsymbol{x}_0 - \boldsymbol{x}^*\|_2^2, \tag{3}$$

where x_0 denotes the initial approximation, assuming that $\mathrm{A}\boldsymbol{x} = \boldsymbol{b}$ has a solution and denote $\mathbf{x}^* := \mathrm{A}^\dagger \mathbf{b}$, $\mathrm{R} = \|\mathrm{A}^{-1}\|_2^2 \|\mathrm{A}\|_F^2$, $\|\mathrm{A}^{-1}\|_2$ is the smallest constant M such that the inequality $\|\mathrm{A}\boldsymbol{x}\|_2 \geq \frac{1}{\mathrm{M}}\|\boldsymbol{x}\|_2$ holds for all vectors x. The RK method addresses the significant disadvantage of the original Kaczmarz algorithm. It is easy to analyze the convergence rate for the RK method, while the original algorithm converge very slow when the data order is poor.

In 2010, D. Needell [6] analyzed the case where the system $\mathrm{A}\boldsymbol{x} = \boldsymbol{b}$ is corrupted by noise, and considered the system $\mathrm{A}\boldsymbol{x} = \boldsymbol{b} + \boldsymbol{r}$ where r is an arbitrary error vector. In the noisy version, she obtained the RK method reaches an error threshold dependent on the matrix A with the same rate as in the error-free case,

$$\mathbb{E}\|\boldsymbol{x}_k^* - \boldsymbol{x}^*\|_2 \leq \left(1 - \frac{1}{\mathrm{R}}\right)^{\frac{k}{2}} \|\boldsymbol{x}_0 - \boldsymbol{x}^*\|_2 + \sqrt{\mathrm{R}}\gamma, \tag{4}$$

where $\gamma = \max_i |r_i| / \|\boldsymbol{a}_i\|_2$, x_k^* denote the kth iterate of noisy randomized Kacmarz. Later, in[7], D. Needell and Y. C. Eldar utilized the Johnson-Lindenstrauss[7] dimension reduction technique to keep the runtime on the same order as the original RK version and proposed randomized Kaczmarz via Johnson-Lindenstrauss(RKJL) and improved the convergence rate of the RK method.

In [8], A. Zouzias and N. M. Freris presented the randomized extended Kaczmarz for solving inconsistent linear system $\mathrm{A}\boldsymbol{x} = \boldsymbol{b} + \boldsymbol{r}$. By a randomized approximation orthogonal projection, they approximately computed linear equation of $\mathrm{A}\boldsymbol{x} = \boldsymbol{b}_{\mathrm{R(A)}}$, where $\mathcal{R}(\mathrm{A}) := \{\mathrm{A}\boldsymbol{x} \mid \boldsymbol{x} \in \mathbb{R}^n\}$. Essentially, the RK algorithm is applied twice. The convergence rate of this method is

$$\mathrm{E}\|\boldsymbol{x}_k - \boldsymbol{x}^*\|_2^2 \leq \left(1 - \frac{1}{\mathrm{R}}\right)^{\left\lfloor \frac{k}{2} \right\rfloor} \left(1 + 2\mathrm{R}_1^2\right) \|\boldsymbol{x}_0 - \boldsymbol{x}^*\|_2^2, \tag{5}$$

Let $\sigma_1 \geq \sigma_2 \geq \ldots \geq \sigma_{rank(A)}$ be the non-zero singular values of A, where R_1^2 denotes $\sigma_1^1 / \sigma_{rank(A)}^2$. This method can be considered as a randomized variant of the extended Kaczmarz method proposed by C.Popa [9].

In [10], J. Liu and S. J. Wright proposed an accelerated randomized(ARK) algorithm with better convergence than the standard RK method on ill conditioned problems. The ARK method starts at a significant disadvantage in the sparse setting, so by Nesterov's accelerated procedure [11], they improved the iteration of the standard RK method and proved, if matrix A is dense and the minimum singular value of $A^T A$ is small,

$$\mathrm{E}\|\boldsymbol{x}_k - \boldsymbol{x}^*\|_2^2 \leq \frac{4\mathrm{m}^2 \|\boldsymbol{x}_0 - \boldsymbol{x}^*\|_{(\mathrm{A^T A})^+}^2}{\mathrm{k}^2}, \tag{6}$$

where $\|\boldsymbol{x}_0 - \boldsymbol{x}^*\|_{(\mathrm{A^T A})^+}^2$ is $\sqrt{\mathrm{trace}((\boldsymbol{x}_0 - \boldsymbol{x}^*)^{\mathrm{T}} (\mathrm{A^T A})^+ (\boldsymbol{x}_0 - \boldsymbol{x}^*))}$, $(A^T A)$ is the Moore-Penrose pseudoinverse of $A^T A$. Motivated by the idea of Nesterov's acceleration, unlike the asynchronous parallel randomized Kaczmarz algorithm, we apply Nesterov's accelerated procedure in the REK method if A is dense and the minimum singular value of $A^T A$ is small, and propose an accelerated randomized extended Kaczmarz algorithm(AREK). In this paper, under the condition that A is dense and the minimum singular value of $A^T A$ is small, we prove that the convergence rate of the AREK method is better than the REK method, and illustrate the computational results of our approach.

2. Algorithm

Both theoretical and numerical researches have shown that the RK algorithm provides very promising results. Here we mainly show an accelerated REK method works well in the case where the system is corrupted with noise. In this section we consider the consistent system $\mathrm{A}\boldsymbol{x} = \boldsymbol{b}$ after an error r is added to the right side, $\mathrm{A}\boldsymbol{x} \approx \boldsymbol{b} + \boldsymbol{r}$ In (4) with $\boldsymbol{r} = \boldsymbol{b}_{\mathcal{R}(\mathrm{A})}$ tells us that the randomized Kaczmarz algorithm performs well for least square problems whose least squares error is very close to zero, $\| \boldsymbol{r} \|_2 \approx 0$. First we give the notations and preliminaries.

2.1 Notations and preliminaries

We denote the rows and columns of A by $\boldsymbol{a}_1,...\boldsymbol{a}_m$ and $\boldsymbol{a}^1,...,\boldsymbol{a}^n$. For an integer m, let $[m] := \{1,2,...,m\}$. Let $\sigma_1 \geq \sigma_2 \geq ... \geq \sigma_{rank(A)}$ be the non-zero singular values of A, $\sigma_1 = \sigma_{\min}$ and $\sigma_{rank(A)} = \sigma_{\max}$. $\mathcal{R}(A) := \{A\boldsymbol{x} \mid \boldsymbol{x} \in \mathbb{R}^n\}$ and $\mathcal{R}(A)^{\perp}$ denotes the orthogonal complement of $\mathcal{R}(A)$. Assume that $A\boldsymbol{x} = \boldsymbol{b}$ has a solution and denote $\mathbf{x}^* := A^{\dagger}\mathbf{b}$, and the fist estimate $\boldsymbol{x}_0 = 0$. Given any $\boldsymbol{b} \in \mathbb{R}^m$, we can write it as $\boldsymbol{b}_{\mathcal{R}(A)} + \boldsymbol{b}_{\mathcal{R}(A)\perp}$, where $\boldsymbol{b}_{\mathcal{R}(A)}$ is the projection of b onto $\mathcal{R}(A)$. $\| A \|_F$ is the Frobenius norm of A. Given a positive semidefinite matrix M, $\|X\|_M$ is defined as $\sqrt{trace(X^T M X)}$. $R = \|A^{-1}\|_2^2 \|A\|_F^2$, $\|A^{-1}\|_2$ is the smallest constant M such that the inequality $\|A\boldsymbol{x}\|_2 \geq (1/M)\|\boldsymbol{x}\|_2$ holds for all vectors $\boldsymbol{x}$. We assume throughout that the rows of A are normalized, $\|\boldsymbol{a}_i\|_2 = 1$, $i = 1,2,...,m$.

2.2 Randomized extended Kaczmarz(REK)

Table 1. Randomized extended Kaczmarz(REK) (Algorithm 1)

Require:	$m \times n$ matrix A, coefficient vector $\boldsymbol{b} \in \mathbb{R}^m$
Initialization:	$\boldsymbol{x}_0 = 0$ and $z_0 = \boldsymbol{b}$.
Select:	Pick $i_k \in [m]$, $j_k \in [n]$ with probability $q_i := \frac{\|\boldsymbol{a}_i\|_2^2}{\|A\|_F^2}, i \in [m]$ and $p_j := \frac{\|\boldsymbol{a}^j\|_2^2}{\|A\|_F^2}, j \in [n]$
Project:	$z_{k+1} = z_k - \frac{\langle \boldsymbol{a}^{j_k}, z_k \rangle}{\|\boldsymbol{a}^{j_k}\|_2^2} \boldsymbol{a}^{j_k}$, $\boldsymbol{x}_{k+1} = \boldsymbol{x}_k + \frac{b^{i_k} - z_k^{i_k} - \langle \boldsymbol{x}_k, \boldsymbol{a}_{i_k} \rangle}{\|\boldsymbol{a}_{i_k}\|_2^2} \boldsymbol{a}_{i_k}$
Update:	Set k = k + 1
end	

A. Zouzias and N. M. Freris proposed the variant of the RK method for solving inconsistent linear system, see [8] for more details.
This approach can efficiently reduce the norm of the noisy part of b, then apply the RK algorithm on a new linear system whose right hand side vector is now arbitrarily close to the column space of A, $A\boldsymbol{x} \approx \boldsymbol{b}_{\mathcal{R}(A)}$. Algorithm 1, see Table 1, showed that the RK method is applied twice by the REK method, and The convergence rate of it has been proven, it is showed in (5). Actually, A. Zouzias and N. M. Freris did not give the best result of the convergence rate of the REK method. By the method of their proof, it is easy to obtain the convergence rate of the REK algorithm,

Theorem 1. After $k > 1$ iterations, let $\boldsymbol{x}_k^*$ be the kth iterate of the noisy REK method, then,

$$E\|\boldsymbol{x}_k^* - \boldsymbol{x}^*\|_2^2 \leq (1 - \frac{1}{R})^k \|\boldsymbol{x}_0 - \boldsymbol{x}^*\|_2^2 + 2\left(1 - \frac{1}{R}\right)^{\left\lfloor \frac{k}{2} \right\rfloor} \|\boldsymbol{x}_0 - \boldsymbol{x}^*\|_2^2 \frac{\sigma_{\max}^2(A)}{\sigma_{\min}^2(A)}, \tag{7}$$

where $R = \|A^{-1}\|_2^2 \|A\|_F^2$, and the expectation is taken over the choice of the rows in the algorithm.

We now describe the complexity of the REK method for the case of dense A. In Section 2, We assume throughout that the rows of A are normalized, $\|\boldsymbol{a}_i\|_2 = 1, i = 1,2,...,m$. Then the computation of Algorithm 1 requires about 4(m + n) operations per iteration.

2.3 *Nesterov's Accelerated Randomized Kaczmarz(ARK)*

Table 2. Accelerated Randomized Kaczmarz(ARK) (Algorithm 2)

Require:	$m\times n$ matrix A, coefficient vector $\boldsymbol{b}\in\mathbb{R}^m$
Initialization:	$\boldsymbol{v}_0 = \boldsymbol{x}_0 = 0$. Choose γ_K to be the larger root as in (8). Set α_k and β_k as in (9).
Select:	Pick $i_k \in [m]$ with probability $q_i := \dfrac{\|\boldsymbol{a}_i\|_2^2}{\|A\|_F^2}, i\in[m]$
Project:	$\boldsymbol{y}_k = \alpha_k \boldsymbol{v}_k + (1-\alpha_k)\boldsymbol{x}_k; \boldsymbol{x}_{k+1} = \boldsymbol{y}_k - \boldsymbol{a}_{i_k}(\boldsymbol{a}_{i_k}^T y_k - \boldsymbol{b}^{i_k})/\|\boldsymbol{a}_{i_k}\|_2^2;$ $\boldsymbol{v}_{k+1} = \boldsymbol{\beta}_k \boldsymbol{v}_k + (1-\beta_k)\boldsymbol{y}_k - \gamma_k \boldsymbol{a}_{i_k}(\boldsymbol{a}_{i_k}^T \boldsymbol{y}_k - \boldsymbol{b}^{i_k})/\|\boldsymbol{a}_{i_k}\|_2^2$
Update:	Set k = k + 1
end	

The ARK algorithm, see Table 2, applies Nesterov's accelerated procedure [11] to improve the standard RK method. If we apply $A\boldsymbol{x} = \boldsymbol{b}$ to $\min_x \|A\boldsymbol{x} - \boldsymbol{b}\|_2$, gradient descent sets $\boldsymbol{x}_{k+1} \leftarrow \boldsymbol{x}_k - \theta_k \nabla \boldsymbol{f}(\boldsymbol{x}_k)$, where $\nabla \boldsymbol{f}(\boldsymbol{x}_k)$ is the objective gradient and θ_k is the stepsize. Nesterov's accelerated procedure introduces two sequences $\{\boldsymbol{y}_k\}$ and $\{\boldsymbol{v}_k\}$ and obtain the following iterative scheme: $\boldsymbol{y}_k \leftarrow \alpha_k \boldsymbol{v}_k + (1-\alpha_k)\boldsymbol{x}_k; \boldsymbol{x}_{k+1} \leftarrow \boldsymbol{y}_k - \theta_k \nabla \boldsymbol{f}(\boldsymbol{y}_k); \boldsymbol{v}_{k+1} \leftarrow \beta_k \boldsymbol{v}_k + (1-\beta_k)\boldsymbol{y}_k + \gamma_k \nabla \boldsymbol{f}(\boldsymbol{y}_k)$. With appropriate choice of α_k , β_k and γ_k, this method yields better convergence rate than the standard gradient descent. And in [12], Y. Nesterov obtained the way of choosing the parameter α_k , β_k and γ_k , we set $\gamma_{-1} = 0$, choose γ_k to be the larger root of

$$\gamma_k^2 - \frac{\gamma_k}{m} = (1 - \frac{\gamma_k \lambda}{m})\gamma_{k-1}^2, \tag{8}$$

Then we set α_k and β_k as follows:

$$\alpha_k = \frac{m - \gamma_k \lambda}{\gamma_k (m^2 - \lambda)}, \ \beta_k = 1 - \frac{\gamma_k \lambda}{m}. \tag{9}$$

In [10], J. Liu and S. J. Wright showed the ARK method to address the consistent linear system.Since the main computation of REK is about 4m+4n operations per iteration then the cost per iteration of Algorithm 2 is about 11n with the same assumption as the REK method, incurred in steps of Nesterov's accelerated procedure.

Remark 1. If the data matrix A is sparse, with a fraction of δ nonzeros(with $0 < \delta \le 1$), the ARK method has s significant disadvantage. The average number of operations for each iteration of the RK method is $O(\delta n)$, however, since the vectors $\boldsymbol{x}_k$, $\boldsymbol{y}_k$ and $\boldsymbol{v}_k$ are dense in general, the operation of the ARK algorithm remains at $O(n)$. Therefore, in this paper, we assume that the data matrix A is dense.

2.4 *Nesterov's Accelerated Randomized extended Kaczmarz(AREK)*

We present Nesterov's accelerated randomized extended Kaczmarz algorithm which is a specific combination of the REK algorithm together with the ARK algorithm. The proposed algorithm consists of two components. The first component is the randomized approximate orthogonal projection which implicitly maintains an approximation to $\boldsymbol{b}_{\mathcal{R}(A)}$ formed by $\boldsymbol{b}-z_k$. The second part applies the ARK method, with Nesterov's accelerated procedure, on the system $Ax=b-z_k$.

Table 3. Nesterov's Accelerated Randomized extended Kaczmarz(AREK) (Algorithm 3)

Require:	$m\times n$ matrix A, coefficient vector $b\in\mathbb{R}^m$
Initialization:	$\boldsymbol{v}_0=\boldsymbol{x}_0=0, z_0=\boldsymbol{b}, \gamma_{-1}=0.$
Select 1:	Pick $j_k\in[n]$ with probability $\mathrm{p_j}:=\frac{\lVert \boldsymbol{a}^{\mathrm{j}}\rVert_2^2}{\lVert \mathrm{A}\rVert_{\mathrm{F}}^2}, \mathrm{j}\in[\mathrm{n}]$
Project 1:	Set $\boldsymbol{z}_{\mathrm{k+1}}=\boldsymbol{z}_{\mathrm{k}}-\frac{\langle \boldsymbol{a}^{\mathrm{jk}},\boldsymbol{z}_{\mathrm{k}}\rangle}{\lVert \boldsymbol{a}^{\mathrm{jk}}\rVert_2^2}\boldsymbol{a}^{\mathrm{jk}}$. Choose γ_k to be the larger root as in (8). Set α_k and β_k as in (9).
Select 2:	Pick $i_k\in[m]$ with probability $\mathrm{q_i}:=\frac{\lVert \boldsymbol{a}_{\mathrm{i}}\rVert_2^2}{\lVert \mathrm{A}\rVert_{\mathrm{F}}^2}, \mathrm{i}\in[\mathrm{m}]$
Project 2:	$\boldsymbol{y}_{\mathrm{k}}=\alpha_{\mathrm{k}}\boldsymbol{v}_{\mathrm{k}}+(1-\alpha_{\mathrm{k}})\boldsymbol{x}_{\mathrm{k}};\boldsymbol{x}_{\mathrm{k+1}}=\boldsymbol{y}_{\mathrm{k}}-\boldsymbol{a}_{\mathrm{i_k}}(\boldsymbol{a}_{\mathrm{i_k}}^{\mathrm{T}}\boldsymbol{y}_{\mathrm{k}}-(\boldsymbol{b}^{\mathrm{ik}}-z_{\mathrm{k}}^{\mathrm{ik}}))/\lVert \boldsymbol{a}_{\mathrm{i_k}}\rVert_2^2;$ $\boldsymbol{v}_{\mathrm{k+1}}=\beta_{\mathrm{k}}\boldsymbol{v}_{\mathrm{k}}+(1-\beta_{\mathrm{k}})\boldsymbol{y}_{\mathrm{k}}-\gamma_{\mathrm{k}}\boldsymbol{a}_{\mathrm{i_k}}(\boldsymbol{a}_{\mathrm{i_k}}^{\mathrm{T}}\boldsymbol{y}_{\mathrm{k}}-(\boldsymbol{b}^{\mathrm{ik}}-z_{\mathrm{k}}^{\mathrm{ik}}))/\lVert \boldsymbol{a}_{\mathrm{i_k}}\rVert_2^2.$
Update:	Set k = k + 1
end	

The main computation of Algorithm, see Table 3, consists two parts, the first is 4m which is the cost of randomized approximate orthogonal projection, the second component is 11n of that of the ARK method. Next, we consider the expected rate of convergence AREK.

Theorem 2. If data matrix A is dense and $\lambda_{\min}(A^TA)\to 0^+$, after $k>1$ iterations, AREK with input A and $\boldsymbol{b}$ computes $\boldsymbol{x}$ vector $\boldsymbol{x}^*$ such that

$$\mathrm{E}\lVert \boldsymbol{x}_{\mathrm{k}}^*-\boldsymbol{x}^*\rVert_2^2\le\frac{4\mathrm{m}^2\lVert \boldsymbol{x}_0-\boldsymbol{x}^*\rVert_{(\mathrm{A^TA})}^2}{\mathrm{k}^2}+\left(1-\frac{1}{\mathrm{R}}\right)^{\mathrm{k}}\lVert \boldsymbol{x}_0-\boldsymbol{x}^*\rVert_2^2\frac{\sigma_{\max}^2}{\lVert \mathrm{A}\rVert_{\mathrm{F}}^2}. \tag{10}$$

Before proving Theorem 2, it is first really important to analyze that, when the error vector is added, what will happen to the solution space of the original equations. Next lemma will show that each hyperplane, which is $\mathrm{H_i}=\{\boldsymbol{x}:\langle \boldsymbol{a}_{\mathrm{i}},\boldsymbol{x}\rangle=\boldsymbol{b}_{\mathrm{i}}\}$ is shifted, and the distance of each hyperplane.

Lemma 1. [6] $H_i = \{x : \langle a_i, x \rangle = b_i\}$ is the affine subspace of the $\mathbb{R}^n$ which consist the solution of the unperturbed equations, is the solution space of the noisy equations. Then $\mathrm{H}_i^* = \{\omega + \alpha_i \boldsymbol{a}_i : \omega \in \mathrm{H}_i\}$ where $\alpha_i = \frac{r_i}{\|\boldsymbol{a}_i\|_2^2}$.

Then we will utilize the convergence rate of the part of the randomized approximate orthogonal projection to obtain convergence rate of our approach.

Lemma 2. [8] Let $\mathrm{A} \in \mathbb{R}^{m\times n}, \boldsymbol{b} \in \mathbb{R}^m$ be the variable of linear system $\mathrm{A}\boldsymbol{x} = \boldsymbol{b}$, after $k > 0$ iterations of Algorithm 3 it holds that

$$\mathrm{E}\left\|z_k - \boldsymbol{b}_{\mathcal{R}(\mathrm{A})^\perp}\right\|_2^2 \le \left(1 - \frac{1}{\mathrm{R}}\right)^{\boldsymbol{k}} \left\|\boldsymbol{b}_{\mathcal{R}(\mathrm{A})}\right\|_2^2, \tag{11}$$

Proof of Theorem 2. We denote x_k^* be the kth iterate of the noisy AREK method, and x_k is the kth iterate of the noiseless AREK method. Then we have

$$\mathrm{E}\left\|\boldsymbol{x}_k^* - \boldsymbol{x}^*\right\|_2^2 = \mathrm{E}\left\|\boldsymbol{x}_k^* - \boldsymbol{x}_k + \boldsymbol{x}_k - \boldsymbol{x}^*\right\|_2^2 = \mathrm{E}\left\|\boldsymbol{x}_k^* - \boldsymbol{x}_k\right\|_2^2 + \mathrm{E}\left\|\boldsymbol{x}_k - \boldsymbol{x}^*\right\|_2^2.$$

Let y_{k-1}^* denote the $(k-1)^{th}$ iterate of the noisy AREK method. We us the notation as in Lemma 1, H_i^* and H_i are the solution spaces chosen in the kth iteration. Again by orthogonality, we have

$\mathrm{E}\left\|\boldsymbol{x}_k^* - \boldsymbol{x}_k\right\|_2^2 = \mathrm{E}\left\|\alpha_i \boldsymbol{a}_i\right\|_2^2$. If data matrix A is dense and $\lambda_{\min}(A^T A) \to 0^+$, then by (6), we obtain

$$\mathrm{E}\left\|\boldsymbol{x}_k^* - \boldsymbol{x}^*\right\|_2^2 = \mathrm{E}\left\|\boldsymbol{x}_k - \boldsymbol{x}\right\|_2^2 + \mathrm{E}\left\|\alpha_i \boldsymbol{a}_i\right\|_2^2 \le \frac{4\mathrm{m}^2 \left\|\boldsymbol{x}_0 - \boldsymbol{x}^*\right\|_{(\mathrm{A}^T\mathrm{A})^+}^2}{\mathrm{k}^2} + \mathrm{E}\left\|\alpha_i \boldsymbol{a}_i\right\|_2^2.$$

Since $\alpha_i = \frac{r_i}{\|\boldsymbol{a}_i\|_2^2}$ and $\|\boldsymbol{a}_i\|_2 = 1$, we have $\mathrm{E}\left\|\alpha_i \boldsymbol{a}_i\right\|_2^2 = \frac{\|\boldsymbol{r}\|_2^2}{\|\mathrm{A}\|_F^2}$.

After k iterations, we obtain $\boldsymbol{b} - z_k \approx \boldsymbol{b}_{\mathcal{R}(\mathrm{A})}$, then we have $\boldsymbol{r} = z_k - \boldsymbol{b}_{\mathcal{R}(\mathrm{A})^\perp}$. By Lemma 2,

$$\frac{\|\mathrm{r}\|_2^2}{\|\mathrm{A}\|_\mathrm{F}^2} = \frac{\left\|\boldsymbol{b}_{\mathrm{R(A)}^\perp} - z_k\right\|_2^2}{\|\mathrm{A}\|_\mathrm{F}^2} \le \left(1 - \frac{1}{\mathrm{R}}\right)^{\mathrm{k}} \frac{\left\|\boldsymbol{b}_{\mathrm{R(A)}}\right\|_2^2}{\|\mathrm{A}\|_\mathrm{F}^2}.$$ Since $\left\|\boldsymbol{b}_{\mathcal{R}(\mathrm{A})}\right\|_2 \le \sigma_{\max}\left\|\boldsymbol{x}_0 - \boldsymbol{x}^*\right\|_2$, we obtain (10).

If data matrix A is dense and $\lambda_{\min}(A^T A) \to 0^+$, by comparing the conclusion of Theorem 1 and Theorem 2, we conclude that our approach is faster than the REK method.

3. Numerical results

In the section, we describe some numerical results for the REK algorithm and the AREK algorithm. We compare REK and AREK for dense A and $\lambda_{\min}(A^T A) \to 0^+$. Synthetic data for our numerical computation is generated as follows: optimal solution $\boldsymbol{x}^* \in \mathbb{R}^n$ is chosen to be i.i.d. $N(0,1)$. Each row of A are normalized. We set the right-hand side $\boldsymbol{b} = A\boldsymbol{x}^* + \boldsymbol{r}$, and we set r i.i.d. $(1/100)N(0,1)$ and the data matrix $A \in \mathbb{R}^{m \times n}$ i.i.d. $(1/10)N(0,1)$. We measure performance by plotting residual $\left\|\boldsymbol{x}_0 - \boldsymbol{x}^*\right\|_2$ against the number of iterations, and the initial point $\boldsymbol{x}_0 = 0$ is used in all numerical computation. Figure 1 show residual errors for REK and AREK with different $\lambda_{\min}(A^T A)$ and n.

4. Conclusion

This paper proposes an accelerated randomized extended Kaczmarz(AREK) algorithm via Nesterov's accelerated procedure, and we obtain better convergence rate. Both REK and AREK have almost the same complexities per iteration for the dense data. In the section of numerical results, under the assumption of dense data matrix A and $\lambda_{\min}(A^T A) \to 0^+$, a comparison of convergence rate of REK and AREK is provided.

5. References

[1] S Kaczmarz 1937 Angenaherte auflsung von system linearer gleichungen *Bulletin International de l0Acadmie Polonaise des Sciences et des Letter*, vol. 35, pp. 355-357

[2] H G Feichtinger, C Cenker, M Mayer, H Steier and T Strohmer 1992 New Variants of the POCS Method using Affine Subspaces of Finite Codimension with Applications to Irregular Sampling *Proceedings of SPIE*, 1818:299-310

[3] F Natterer 2002 The Mathematics of Computerized Tomography *Medical Physics*, 29(1):107-108

[4] T Strohmer, R Vershynin 2006 A randomized solver for linear system with exponential Convergence *In RANDOM 2006 (10th International Workshop on Randomization and Computation)*, number 4110 in Lecture Notes in Computer Science, pp. 499-507.

[5] T Strohmer, R Vershynin 2009 A randomized Kaczmarz algorithm with exponential Convergence *J. Fourier Anal. Appl.*, 15:262-278

[6] D Needell 2010 Randomized Kaczmarz solver for noisy linear system *BIT Numerical Mathematics*, vol. 50, no. 2, pp. 1422-1436

[7] Y C Eldar and D Needell 2011 Acceleration of randomized Kaczmarz method via the JohnsonLindenstrauss lemma *Numerical Algorithms*, vol. 58, no. 2, pp. 163-177

[8] A Zouzias and N M Freris 2012 Randomized extended Kaczmarz for solving least-squares *arXiv:*1205.5770v2

[9] C Popa 1999 Characterization of the solution set of least-squares problems by an extension of Kaczmarz's projection method *Journal of Applied Mathematics and Computing*, vol. 6, pp. 51-64

[10] J Liu and S J Wright 2013 An Accelerated Randomized Kaczmarz Algorithm *arXiv:*1310.2887v1

[11] Y Nesterov 2004 Introductory Lectures on Convex Optimization: A Basic Course *Kluwer Academic Publishers*

[12] Y Nesterov 2012 Efficiency of coordinate descent methods on huge-scale optimization Problems *SIAM Journal on Optimization*, vol. 22, no. 2, pp. 341-362

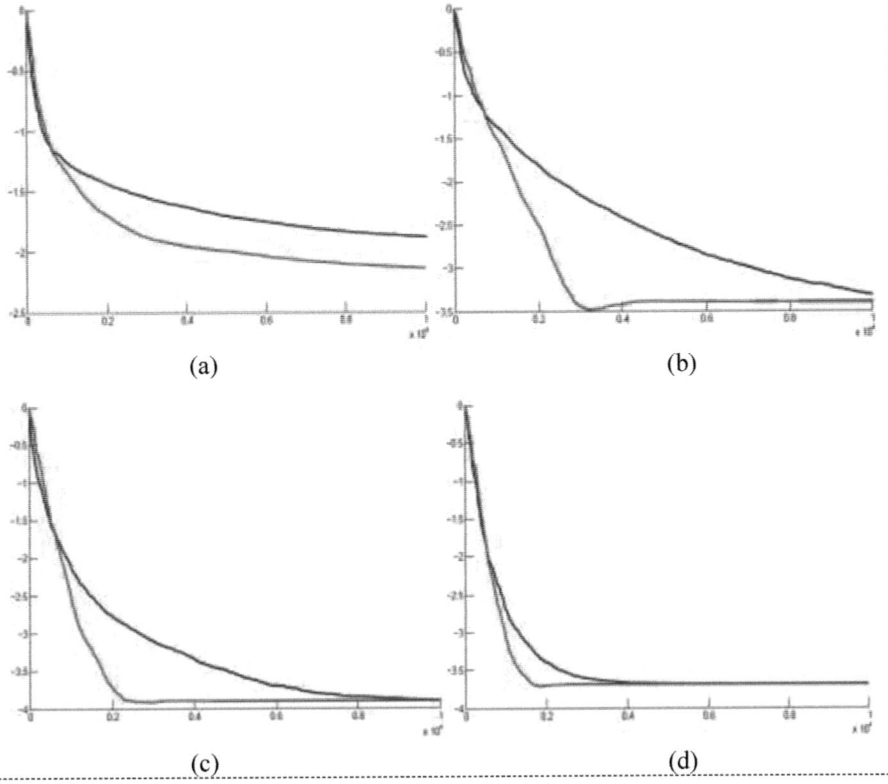

Figure 1. y-axis denotes $\log_{10}\|x_0 - x^*\|_2$, x-axis denotes the number of iterations. Blue line is REK, red line is AREK. (a):m=100, n=90, $\lambda_{\min}(A^T A) = 0.0018$. (b):m=100, n=80, $\lambda_{\min}(A^T A) = 0.0143$. (c):m=100, n=70, $\lambda_{\min}(A^T A) = 0.0267$. (d):m=100, n=60, $\lambda_{\min}(A^T A) = 0.0644$.

Economic growth and concentrations

Shu-hen Chiang

Department of Finance, Chung-Yuan Christian University, Taoyuan, Taiwan

shchiang@cycu.edu.tw

Abstract. Over the past few decades, shift-share (SS) analysis is widely applied to explore the sources of local economic growth; however, it leaves unanswered the inequality question. The purpose of this paper is to exclude these biases caused by inequalities to generate a new identity, which fully shows the concept of externalities and comparative advantage, the nation-industry-region interactions and the structural change of local industry in a timely manner

1. Introduction

Over the past decades, China has been experienced the fastest-growing economy with average annual growth rate of 10% in the world, while China is suffered from the most serious regional inequality of all countries [1, 2, 3, 4]. The conflict between growth and inequality is just a unique China's phenomenon due to the urban-biased policies which pour the huge amounts of social resources into these cities as growth poles that allow China to catch up with the developed economies. Moreover, China's development follows many critical contributions to regional science, for example, growth pole theory and agglomeration economies to propose the concept of regional or industrial clustering as an engine of economic growth and innovation [5, 6]. and these arguments further deteriorate sharply China's inequality problem. In sum, economic polarization in China is actually reflected in imbalance in industrial structure and regional disparities and the inequality question has therefore provoked a great deal of controversy among the authorities, society and researchers in China.

Little attention has been given to a new debate whether inequality question can distort shift-share (SS) analysis, namely, the decomposition of regional economic growth into national, industrial, regional effects externally as well as local specialization effect from comparative advantage. For example, regional or industrial effect of SS analysis is calculated by comparing a region (industry) with its nation based on a classical assumption from regional science that any region or industry is always trivial from macroeconomic angle; that is to say, a region or industry is approximately independent of its nation. However, the fact that regional and industrial inequalities will lead to dominant regions or industries as a critical part of the nation must violate this hypothesis. To make matters worse, when we omit the inequality question to generate three biased external causes, local specialization effect will eventually be incompatible with the law of comparative advantage in need of independencies among different units. Finally, the globalization promotes a higher degree of inequality in order to advance in competitive abilities. It is thus expected that a new decomposition of local growth associated with regional and industrial inequalities is indispensable to SS analysis, especially China.

Even SS analysis is a complicated decomposition approach, the concept of a set is regarded as the best solution: nation can be divided into four separate parts according to industrial and regional classifications and then we can rewrite a new identity with these four areas in order to meet the independent requirements for comparative advantage. At the same time, our new identity possesses desired properties, including being easy to understand, providing a full description of nation-industry-region interaction and searching a correct turning point of industrial transformation. The remainder of this paper is organized as follows. In Section 2, this new version is created to fully remove the impacts of regional and industrial inequalities; at the same time, we discuss some economic meanings and properties of this newly-revised identity. Finally, some conclusions are given in Section 4.

2. A new identity

In this section, we will outline all aspects of deriving a new version, especially based on the concept of a set. In addition, the correctness of this new identity will be later proved and analyzed according to the concept of externalities and comparative advantage.

2.1 Traditional decomposition and the inequality problems

SS analysis, using the method followed by [7] in the context of output may be described as follows:

$$\begin{aligned}\frac{\Delta y_{ij}}{y_{ij}} &= \frac{\Delta y}{y} + \left(\frac{\Delta y_j}{y_j} - \frac{\Delta y}{y}\right) + \left(\frac{\Delta y_i}{y_i} - \frac{\Delta y}{y}\right) + \left(\frac{\Delta y_{ij}}{y_{ij}} - \frac{\Delta y_i}{y_i} - \frac{\Delta y_j}{y_j} + \frac{\Delta y}{y}\right), \\ &= NE + IM + RE + SE\end{aligned} \tag{1}$$

where y_{ij}, y_i, y_j *and* y represent the output level of industry j in region i, the regional output level, the industrial output level and national output with m regions and n industries, respectively. It is clear that (1) is a mathematical identity derived by decomposing the change in industrial output in a region into four portions: NE, IM, RE and SE, which is used to evaluate the growth of local specialization from location quotient (LQ) as the degree of local comparative advantage, which is defined by a ratio of industrial shares in two independent regions as follows:

$$\frac{\Delta LQ_{ij}}{LQ_{ij}} \approx \Delta \ln LQ_{ij} = \Delta \ln\left(\frac{\frac{y_{ij}}{y_i}}{\frac{y_j}{y}}\right) = \frac{\Delta y_{ij}}{y_{ij}} - \frac{\Delta y_i}{y_i} - \frac{\Delta y_j}{y_j} + \frac{\Delta y}{y} = SE, \tag{2}$$

Although the fact that SS analysis is to decompose the change in local output into national, industrial, regional and localized specialization effects seems very reasonable, many overlapping (non-independent) conditions caused by regional and industrial inequalities may give rise to new problems for these four effects. For example, the national economy is an aggregation of regional or industrial elements in (3). By analogy, industrial and regional output variables can be obtained by aggregating industrial output in all regions in (4) and the sum of regional output in all industries in (5).

$$y = \sum_{i=1}^{m} y_i = \sum_{j=1}^{n} y_j = \sum_{i=1}^{m}\sum_{j=1}^{n} y_{ij}, \tag{3}$$

$$y_j = \sum_{i=1}^{m} y_{ij}, \tag{4}$$

$$y_i = \sum_{j=1}^{n} y_{ij}, \tag{5}$$

Judged from the above equations (3) to (5), comparison between a region or industry and its nation is valid only under the assumption that all inequality possibilities can be ignored based on four restrictions: $\frac{y_{ij}}{y_i} = \frac{y_{ij}}{y_j} = \frac{y_i}{y} = \frac{y_j}{y} \approx 0$. Among these situations, $\frac{y_{ij}}{y_i} = \frac{y_j}{y} \approx 0$ implies that the shares of industry j in region i and its nation both must be very small; at the same time, regional distributions in this nation and industry j are both much more diversified for $\frac{y_i}{y} = \frac{y_{ij}}{y_j} \approx 0$. However, these conditions completely conflict with the regional and industrial inequalities, which are caused by industrial and spatial concentrations based on agglomeration economies and growth pole theory as the critical conceptions in the field of regional economics. All of the above makes it clear that omitting the inequality dispute is unsatisfactory for correctly decomposing every source of China's growth.

2.2 The concept of a set

As can be seen from the above, industrial and spatial dimensions each make the inequality issue more complex. The concept of a set is proposed in order to resolve all questions simultaneously. The combination of all types of overlapping opportunities as shown in (3), (4) and (5) can be expressed by using a diagram as shown in figure 1 for the output of nth industry in m^{th} region. This provides a convenient way of setting up a truly separated environment to exclude all possible overlapping channels caused by inequalities.

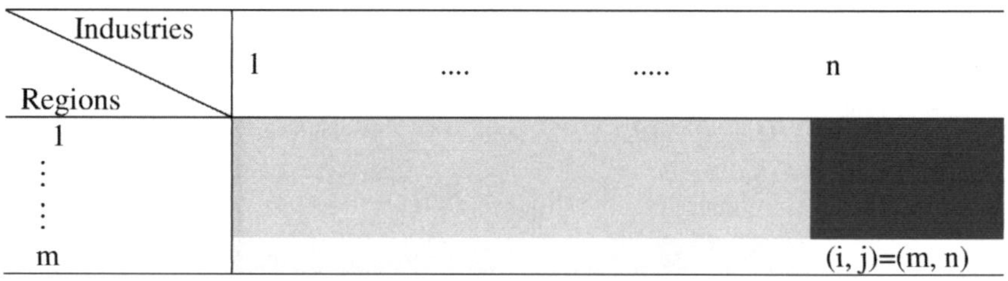

Figure 1. Industries and regions in a nation

There are many important implications in figure 1. First, to obtain the new macroeconomic variable, we must first deduct specific regional and industrial factors with the exception of y_{ij}, namely, the union of two sets (industrial and regional dimensions). In other words, the new national variable ($y' = y - y_j - y_i + y_{ij}$) as the green zone is totally different from original national factor (y) and this variable is defined by the output of other industries in other regions. Secondly, the new industrial (regional) variable, namely, $y_j' = y_j - y_{ij}$ ($y_i' = y_i - y_{ij}$) as the yellow and blue zones, respectively, is a specific industrial (regional) output minus output of industry j in region i. Thus, the new industrial and regional variables correspond to industry j in other regions and other industries in region i, respectively. As far as industry j in region i is concerned, it remains the same as the white zone. Finally, as can be seen in figure 1, it reveals that this new version is especially noteworthy in the case of a dominant region or industry on the grounds that this region or industry caused by regional or industrial inequalities occupies a major area, compared to its nation. It is obvious that original assumption is problematic. In summary, the sum of the four zones is equal to the national economy. This meets the requirement of the separation theorem and we can prove that the four variables must be independent of each other.

2.3 New version and LQ

As soon as the adjusted nation, industry and region, which are added to the local industry, are all determined independently, we can inquire into this new identity through SS analysis as follows:

$$\frac{\Delta y_{ij}}{y_{ij}}=\frac{\Delta y'}{y'}+\left(\frac{\Delta y_j'}{y_j'}-\frac{\Delta y'}{y'}\right)+\left(\frac{\Delta y_i'}{y_i'}-\frac{\Delta y'}{y'}\right)+\left(\frac{\Delta y_{ij}}{y_{ij}}-\frac{\Delta y_i'}{y_i'}-\frac{\Delta y_j'}{y_j'}+\frac{\Delta y'}{y'}\right)$$
$$=RNE+RIM+RRE+RSE \qquad (6)$$

For convenience of analysis, our identity in (6) is separated into four new effects: revised national effect (RNE), revised industrial mix (RIM), revised regional effect (RRE) and revised specialized effect (RSE) based on four new variables in Section 3.2. To advance the reliability of new version, it is worthwhile examining the relationship between RSE and the revised LQ (RLQ) more closely. Let us start with the definition of the RLQ in independent case from [8]. as shown in (7).

$$RLQ=\frac{y_{ij}\Big/ y_i-y_{ij}}{y_j-y_{ij}\Big/ y-y_j-y_i+y_{ij}} \qquad (7)$$

Through some mathematical operations in RSE as in (8), it is obvious that the RSE actually stems from the RLQ. It is better to say that the link between LQ and SE on classic conditions in (1) and (2) just corresponds to the relationship between RLQ and RSE based on the independent positions in (7) and (8). At the same time, from figure 1, it is clear that RLQ is a ratio of industrial shares in two regions, namely, region i and other regions and this is exactly the definition of comparative advantage. According to the above, we come to the conclusion that this revised version can greatly improve the fundamentals of SS analysis in complete agreement with the concept of comparative advantage.

$$RSE=\left(\frac{\Delta y_{ij}}{y_{ij}}-\frac{\Delta y_i'}{y_i'}-\frac{\Delta y_j'}{y_j'}+\frac{\Delta y'}{y'}\right)=\Delta\ln\left(\frac{y_{ij}\Big/ y_i'}{y_j'\Big/ y'}\right)$$
$$=\Delta\ln\left(\frac{y_{ij}\Big/ y_i-y_{ij}}{y_j-y_{ij}\Big/ y-y_j-y_i+y_{ij}}\right)\approx\frac{\Delta RLQ}{RLQ} \qquad (8)$$

3. Economic meanings and policy

3.1 Comparison between old and new identities

It is believed that comparing the new and old identities can provide additional information to evaluate the merits of this new version. The following serves as an example with regional and industrial inequalities: a high-growth economically dominant area with a high-growth industry which plays a critical role in its nation. How can we decompose local-industry growth into four effects? This case will generate four outcomes: RNE<NE, RIM<IM, RRE<RE and a far higher RSE than SE due to the same magnitude of Δy_{ij}. This result tells us that the change in local competitiveness of this industry can be truly reflected in RSE, while an original SE is the wrong index for assessing local competitiveness.

Moreover, since there are other industries in other regions (y'), this industry in other regions (y'_j) and other industries in this region (y'_i) are all external to this industry in this region (y_{ij}), RNE, RIM and RRE can be regarded as three external sources of growth and change in local industrial output. Finally, RSE, representing the degree of industrial specialization by itself in this region based on the law of comparative advantage, is an internal factor. These outcomes are fully consistent with the spirit of SS analysis, where shift and share represent the changes in any economic variable from an internal factor and three external causes, respectively.

3.2 A manifestation of nation-industry-region interactions

The biggest discrepancy between this new version and the classical identity is our emphasis on the closer and clearer interactions between local industry and other external factors. For example, reference area in this new identity is variable by other industries in other regions as a new national element, while there is the same reference area, namely, nation in a traditional model. In a similar way, reference industry and region both are also flexible according to this industry in other regions and other industries in this region as new industrial and regional elements. In other words, this new identity with variable effects is very reasonable for facing different regions and industries. Finally, this new identity can let us see three types of real interactions: first, nation-local industry interaction is reflected in the relationship between local industry and other industries at other regions as RNE. Second, interregional interactions at the same industry are evaluated by RIM. Third, inter-industrial interactions within a region are appreciated by RRE. The above arguments are very reasonable and intuitive on the grounds that for any local industry, all possible external effects of local industry should inherently be different case by case. In sum, these changing effects of RNE, RIM and RRE can delicately delineate the interactive processes between local industry and three external factors.

4. Conclusion

The neglect of industrial and regional inequality questions in traditional regional science is incompatible with a true economic development like China. Moreover, the arguments regarding spatial and industrial concentrations make the ignorance of the inequality questions more and more preposterous. The purpose of this paper is to generate a new identity through the removal of these possible inequalities to reply to these doubts. Fortunately, the concept of a set is used to divide the national economy into four independent parts and it is proved to be very useful in answering this question. More importantly, our new version shows that local growth depends on three types of externalities as well as local comparative advantage and this outcome is totally consistent with economic theory in general. At the same time, it is believed that the regional and industrial inequality questions may apparently result in an under- or overestimation in all effects and it is confirmed by the well-known example of China with seriously regional and industrial inequalities. Finally, the ability to identify structural change can be greatly advanced using our new version and this is critical for setting a new development strategy in today's globalization era.

In sum, we believe that using this new version can improve the evaluation in the sources of growth and change of the regional economy. It is expected that this study can attract a renewal of interest in interdisciplinary MATEC Web of Conferences studies between regional science and other economic fields.

5. References

[1] Fan, C. C., M. Sun, 2008. Regional inequality in China, 1978-2006. Eurasian Geography and Economics 49, 1-20.
[2] Jones, D. C., L. Cheng, A. L. Owen, 2003. Growth and regional inequality in China during the reform era. China Economic Review 14: 186-200.
[3] Li, H., K. E. Haynes, 2011. Economic structure and regional disparity in China: beyond the Kuznets transition. International Regional Science Review 34: 157-190.
[4] Yang, D. T., 1999. Urban-biased policies and rising income inequality in China. American Economic Review 89, 306-310.

[5] Glaeser E, H. Kallal, J. Scheinkman, A. Schleifer, 1992. Growth in cities. Journal of Political Economy 100: 1126-1152.

[6] Jaffe, A.B., M. Trajtenberg, R. Henderson, 1993. Geographic localization of knowledge spillovers as evidence by patent citations. Quarterly Journal of Economics 108: 577–598.

[7] Coulson, N. E., 1993. The sources of sectoral fluctuations in metropolitan areas. Journal of Urban Economics 33: 76-94

[8] Chiang, S. 2009. Location quotient and trade. Annals of Regional Science 41: 199-214.

Acknowledgements
The authors would like to acknowledge the financial support provided by the Ministry of Science and Technology (MOST) under Grant MOST 104-2410-H-033-048.

IOP Conf. Series: Journal of Physics: Conf. Series **814** (2017) 012019 doi:10.1088/1742-6596/814/1/012019

Approximate solutions for the nonlinear duffing oscillator with an external force

Jutamas Kirinin, Kanokwan Mahithithummath, Narisara Kongsawasdi and Jaipong Kasemsuwan*

Department of Mathematics, Faculty of Scinece, King Mongkut's Institute of Technology Ladkrabang, Thailand

* Corresponding author: kwjaipon@kmitl.ac.th, jaipui@hotmail.com

Abstract. This paper is concerned with a finding of an approximate solution to the Duffing oscillator with a linear external force using a modified differential transform method (MDTM) and Runge-Kutta-Nystrom (RKN) method. The results obtained from the two methods are then numerically and graphically compared. We found that both results are close to one another when an amplitude and frequency of external forces are very small. However, both results are quite different when the amplitude and frequency of external forces are large. Their results are in agreement with the phase plane diagram which demonstrates a stable equilibrium point for both methods in case of small amplitude and frequency of external force. However, for large amplitude and frequency of external forces, only MDTM shows the stable equilibrium point.

1. Introduction

The modified differential transform method (MDTM) is basically based on a conventional DTM [1,2] in combination with the Laplace transform, the Padé approximation and the inverse Laplace transform. It was used to find an approximate solution of nonlinear oscillating equation in [3]. The paper in the reference [4] used the same method to find an approximate solution to the Duffing oscillator with a damping term. Unlike DTM method whose solution is only valid for a short period of time, the results obtained from the MDTM are always valid and stable. In this work, we apply MDTM to find an approximate solution to the Duffing oscillator with a linear external force and compare the results with those obtained from the Runge-Kutta-Nystrom method. The comparisons are in good agreement only when an external force has small amplitude and frequency.

The non-liner Duffing oscillator with an external force and initial conditions in several Physics, Engineering and Mathematics problems [5,6] can be shown as

$$y'' + \delta y' + \alpha y + \beta y^3 = \gamma \cos(\omega t) \quad (1.1)$$

$$y(0) = a, \; y'(0) = b; \qquad a, b \in R \quad (1.2)$$

where δ is a coefficient of delay.
α is a coefficient of stiffness.
β is a coefficient of hardness.
γ is a amplitude of the external force.
ω is a frequency of the external force.

In the next section, we introduce the fundamental definitions and the necessary basic mathematical operations used in the work. In the third section, the derivation procedure of the MDTM for solving the Duffing oscillator with an external force is explained step by step. In the fourth section, many problems with various amplitudes and frequencies of external forces are demonstrated and, finally, the conclusion is provided.

2. Differential transform method (DTM)

Definition 2.1. If $y(t)$ is analytic in time domain T, then it is continuously differentiable in time t. For $t = 0$ and k belong to the set of non-negative integers, denotes as K-domain.

$$Y(k) = \frac{1}{k!}\left[\frac{d^k}{dt^k} y(t)\right]_{t=0}, \forall k \in K, \tag{2.1}$$

where $y(t)$ is the original function and $Y(k)$ is the transform function.

Definition 2.2 The differential inverse transform of $Y(k)$ is defined by power series

$$y(t) = \sum_{k=0}^{\infty} Y(k)t^k. \tag{2.2}$$

From (2.1), the solution of our problem can be expressed by the finite Taylor's series

$$y(t) = \sum_{k=0}^{N}\left[\frac{d^k}{dt^k} y(t)\right]_{t=0} t^k. \tag{2.3}$$

Table 1. Necessary mathematical operator

Original function	Transformed function
$y(t) = \alpha u(t) \pm \beta v(t)$	$Y(k) = \alpha U(k) \pm \beta V(k)$
$y(t) = d^m u(t) / dt^m$	$Y(k) = \frac{(k+m)}{k!} U(k+m)$
$y(t) = u(t)v(t)$	$Y(k) = \sum_{l=0}^{k} U(l)V(k-l)$
$y(t) = t^m$	$Y(k) = (k-m) = \begin{cases} 1, \text{ if } k=m, \\ 0, \text{ if } k \neq m. \end{cases}$
$y(t) = \cos(\omega t + \alpha)$	$Y(k) = (\omega^k / k!) \cos(k\pi / 2 + \alpha)$

3. Using modified differential transform method for solving duffing oscillator with an external force

In this section, we explain the derivation procedure of the MDTM. The problem with initial conditions can be written in the following general form:

$$y'' + \delta y' + \alpha y + \beta y^3 = \gamma \cos(\omega t), \quad y(0) = a, \quad y'(0) = b. \tag{3.1}$$

Applying the differential transforming operators as shown in Table 1 to (3.1), we obtain the transformed recurrence formulas as follows:

$$Y(k+2) = -\frac{1}{(k+1)(k+2)}\left\{\left(\alpha\sum_{k_2=0}^{k}\sum_{k_1=0}^{k_2}Y(k_1)Y(k_2-k_1)Y(k-k_2)\right)+Y(k)+\delta(k+1)Y(k+1)-\left(\gamma\left(\frac{\omega^k}{k!}\right)\cos\left(\frac{k\pi}{2}\right)\right)\right\} \quad (3.2)$$

$$Y(0)=a,\ Y(1)=b. \quad (3.3)$$

From (3.2) and (3.3), the expression of *Y(k)* is obtained. By substituting *Y(k)* to (2.3), the series solution are obtained as

$$y(t) = Y(0)+Y(1)t+Y(2)t^2+Y(2)t^2+Y(3)t^3+\cdots \quad (3.4)$$

Since the differential transform solution in Eq. (3.4) is valid only for a short period of time, the MDTM is used to correct this problem.

Following are the steps in deriving the MDTM:

Step1: Applying the Laplace transform to the series solution (3.4), and setting $s = 1/t$,

Step 2: Applying Padé approximation to the solution from the previous step to obtain the following approximant in the following rational function

$$\left[\frac{L}{M}\right] = \frac{p_0+p_1t+p_2t^2+\cdots+p_Lt^L}{q_0+q_1t+q_2t^2+\cdots+q_Mt^M}$$

and recalling $t = 1/s$,

Step 3: Using the inverse Laplace transform to $[L/M]$ approximant.

4. Case studies and their results

In this section, three examples of Duffing oscillator associated with different external forces are explained. The first example is the Duffing oscillator under a sinusoidal external force. The second example uses the same Duffing oscillator but with a relatively larger amplitude of the external force while the last example use higher frequency of the external force, yet the same amplitude as in the second example. Noted that all of the problems use the same initial conditions for comparison purposes.

Example 4.1

The initial problem of the Duffing oscillator with a small amplitude and frequency of an external force is as follows:

$$y''+0.5y'+25y+25y^3=0.1\cos t \quad (4.1)$$

$$y(0)=0.1,\ y'(0)=0. \quad (4.2)$$

By using the transformed recurrence relation (3.2) and (3.3) when $\delta=0.5, \alpha=25, \beta=25, \gamma=0.1$, $\omega=1$, $a=0.1$ and $b=0$, we obtain

$$Y(k+2) = -\frac{1}{(k+1)(k+2)}\left\{\left(25\sum_{k_2=0}^{k}\sum_{k_1=0}^{k_2}Y(k_1)Y(k_2-k_1)Y(k-k_2)\right)+25Y(k)+0.5(k+1)Y(k+1)-\left(0.1\left(\frac{1^k}{k!}\right)\cos\left(\frac{k\pi}{2}\right)\right)\right\} \quad (4.3)$$

$$Y(0)=0.1,\quad Y(1)=0. \quad (4.4)$$

From (4.3) and (4.4), we obtain *Y(k)* for *k=2,3,…N*. Substituting *Y(k)* to (2.3) and use $N=7$, we obtain the series solution as

$$y(t) = 0.1-1.2125t^2+0.202083t^3+2.572396t^4-0.517422t^5-2.532255t^6+0.585614t^7. \quad (4.5)$$

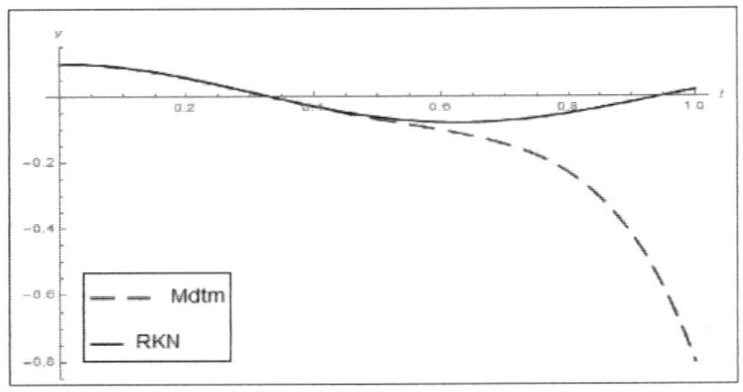

Figure 1. Results from the DTM and RKN for $\gamma = 0.1$, $\omega = 1$.

Figure 1 shows the comparison results obtained by the differential transform method (DTM) (see dashed line) as shown in Eq. (4.5), and the Runge-Kutta-Nystrom (RKN) method (see solid line). It is shown that the results are in good agreement only in the initial stage ($t < 0.6$ s). This error can be improved by using the MDTM via steps previously elaborated.

Step1: Applying the Laplace transform to the series solution (4.5),

$$L\left[y(t)\right] = \frac{2951.492187}{s^8} - \frac{1823.223437}{s^7} - \frac{62.090625}{s^6} + \frac{61.7375}{s^5} + \frac{1.2125}{s^4} - \frac{2.425}{s^3} + \frac{0.1}{s}. \quad (4.6)$$

Setting $s = 1/t$, then

$$L\left[y(t)\right] = 0.1t - 2.425t^3 + 1.2125t^4 + 61.7375t^5 - 62.090625t^6 - 1823.223437t^7 + 2951.492187t^8 \quad (4.7)$$

Step 2: Applying the Padé approximation to the solution (4.7), we obtain the following approximant in the form of rational function

$$\left[\frac{3}{3}\right] = \frac{0.1t + 0.051413t^2 + 0.146583t^3}{1 + 0.514134t + 25.715830t^2 + 0.342756t^3}. \quad (4.8)$$

Recalling that $t = 1/s$, then

$$\left[\frac{3}{3}\right] = \frac{0.146583 + 0.051413s + 0.1s^2}{0.342756 + 25.715830s + 0.514134s^2 + s^3}. \quad (4.9)$$

Step 3: Applying the inverse Laplace transform to the $\left[\frac{3}{3}\right]$ approximant, we then obtain the approximate solution of the problem.

$$y(t) = a_1 e^{-0.013332t} + a_2 e^{(-0.250401-5.064233i)t}((0.995091 + 0.098963i) + a_3 e^{(10.128465i)t}) \quad (4.10)$$

where $a_1 = 0.005677$, $a_2 = (0.047161 - 0.002339i$, $a_3 = (1+i)$

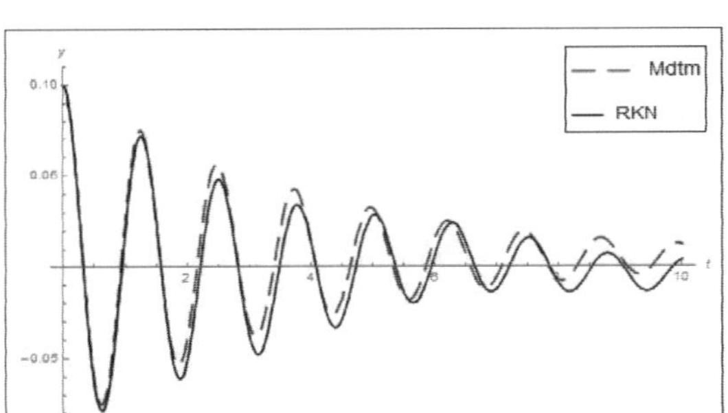

Figure 2. Results from the MDTM and RKN for $\gamma = 0.1$, $\omega = 1$.

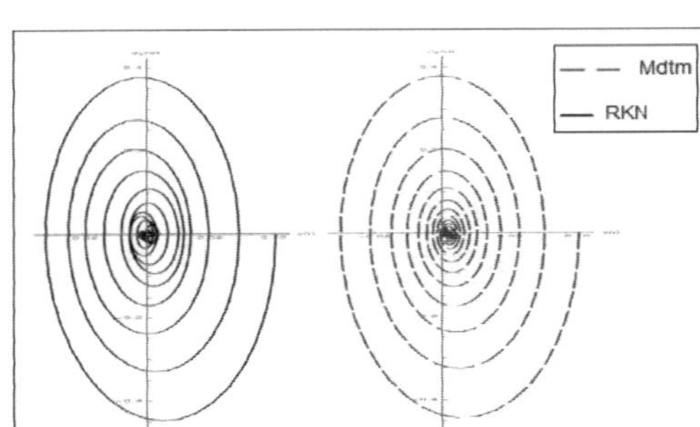

Figure 3. Phase plane diagram from the MDTM and RKN for $\gamma = 0.1$, $\omega = 1$

Figure 2 shows comparison results obtained from the MDTM as shown in Eq. (4.10) and the RKN. It is clearly seen that both results are close to one another for a large time scale.
Figure 3 shows that the Phase plane diagram. Again, it is clearly seen that both results are in good agreement and they demonstrates a stable equilibrium point.

Example 4.2
The initial problem of the Duffing oscillator in this example is the same with the previous one except that the external force has larger amplitude:

$$y''+0.5y'+25y+25y^3=0.5\cos t \tag{4.11}$$
$$y(0)=0.1,\ y'(0)=0. \tag{4.12}$$

By using the transformed recurrence relation (3.2) and (3.3) when $\delta=0.5, \alpha=25, \beta=25, \gamma=0.5$, $\omega=1$, $a=0.1$ and $b=0$, we obtain

$$Y(k+2)=\frac{-1}{(k+1)(k+2)}\left\{\left(25\sum_{k_2=0}^{k}\sum_{k_1=0}^{k_2}Y(k_1)Y(k_2-k_1)Y(k-k_2)\right)+25Y(k)+0.5(k+1)Y(k+1)-\left(0.5\left(\frac{1^k}{k!}\right)\cos\left(\frac{k\pi}{2}\right)\right)\right\} \tag{4.13}$$

$$Y(0)=0.1,\qquad Y(1)=0. \tag{4.14}$$

From (4.13) and (4.14), we obtain *Y(k)* for *k=2,3,...N.* Substituting *Y(k)* to (2.3) and use $N=7$, we obtain the series solution as

$$y(t)=0.1-1.0125t^2+0.168750t^3+2.1307299t^4-0.430339t^5-2.048609t^6+0.471189t^7. \tag{4.15}$$

Step1: Applying the Laplace transform to the series solution (4.15),

$$L\left[y(t)\right]=\frac{2374.792188}{s^8}-\frac{1474.998438}{s^7}-\frac{51.640625}{s^6}+\frac{51.137500}{s^5}+\frac{1.0125}{s^4}-\frac{2.025}{s^3}+\frac{0.1}{s}. \tag{4.16}$$

Setting $s=1/t$, and

Step 2: Applying the Padé approximation to the solution (4.16), we obtain the following approximant in the form of rational function

$$\left[\frac{3}{3}\right]=\frac{0.1t+0.052350t^2+0.526484t^3}{1.0+0.523502t+25.514837t^2+0.4759111t^3}. \tag{4.17}$$

Recalling $t=1/s$, and

Step 3: Applying the inverse Laplace transform to the $\left[\frac{3}{3}\right]$ approximant, we then obtain the approximate solution of the problem.

$$y(t)=a_1e^{-0.018659t}+a_2e^{(-0.252421-5.043977i)t}((0.994811+0.101745i)+a_3e^{(10.087953i)t}) \tag{4.18}$$

where $a_1=0.020612, a_2=(0.039694-0.002025i), a_3=(1+i)$

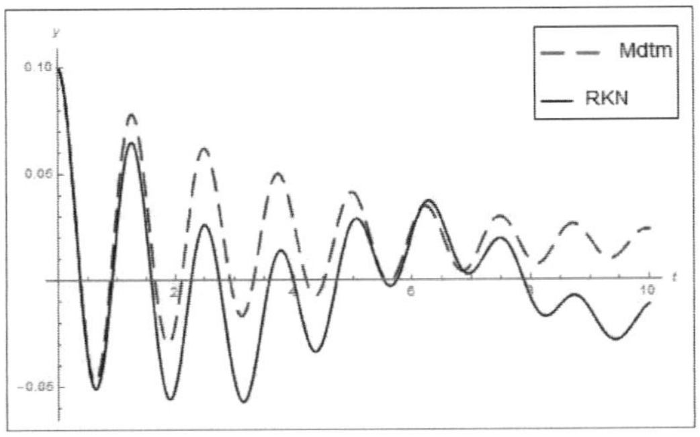

Figure 4. Results from the MDTM and RKN for $\gamma = 0.5,\ \omega = 1$.

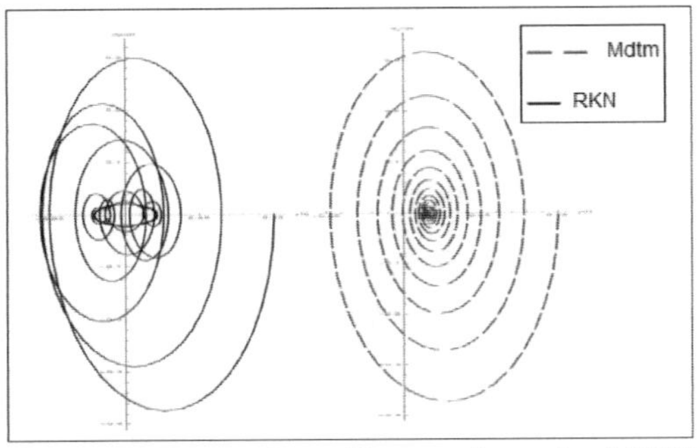

Figure 5. Phase plane diagram from the MDTM and RKN for $\gamma = 0.5,\ \omega = 1$.

Figure 4 shows the comparison results obtained from the MDTM in Eq. (4.10) and form the RKN. It is shown that both results are similar initially but become very different as time increases.
Figure 5 shows the Phase plane diagram. Obviously, the results from RKN shows unstable equilibrium point. However, the results obtained from MDTM shows stable equilibrium point which should be the case since the external is a linear function.

Example 4.3
In this example, the frequency of the external force is set larger as follows:

$$y''+0.5y'+25y+25y^3 = 0.5\cos 3t \tag{4.19}$$

$$y(0) = 0.1 \ ,\ y'(0) = 0. \tag{4.20}$$

By using the transformed recurrence relation (3.2) and (3.3) when $\delta = 0.5, \alpha = 25, \beta = 25, \gamma = 0.5$, $\omega = 2$, $a = 0.1$ and $b = 0$, we obtain

$$Y(k+2) = -\frac{1}{(k+1)(k+2)}\left\{\left(25\sum_{k_2=0}^{k}\sum_{k_1=0}^{k_2} Y(k_1)Y(k_2-k_1)Y(k-k_2)\right)25Y(k)+0.5(k+1)Y(k+1)-\left(0.5\left(\frac{3^k}{k!}\right)\cos\left(\frac{k\pi}{2}\right)\right)\right\} \tag{4.21}$$

$$Y(0) = 0.1, \qquad Y(1) = 0. \tag{4.22}$$

From (4.21) and (4.22), we obtain *Y(k)* for *k=2,3,...N.* Substituting *Y(k)* to (2.3) and use $N = 7$, we obtain the series solution as

$$y(t) = 0.1 - 1.0125t^2 + 0.16875t^3 + 1.96406t^4 - 0.413672t^5 - 1.85139t^6 + 0.446883t^7. \tag{4.23}$$

Step1: Applying the Laplace transform to the series solution (4.23),

$$L[y(t)] = \frac{2252.29}{s^8} - \frac{1333}{s^7} - \frac{49.6406}{s^6} + \frac{47.1374}{s^5} + \frac{1.0125}{s^4} - \frac{2.025}{s^3} + \frac{0.1}{s}. \tag{4.24}$$

Setting $s = 1/t$, and

Step 2: Applying the Padé approximation to the solution (4.24), we obtain the following approximant in the form of rational function

$$\left[\frac{3}{3}\right] = \frac{0.1t + 0.0609443t^2 + 0.456708t^3}{1.0 + 0.609443t + 24.8171t^2 + 2.21621t^3}. \tag{4.25}$$

Recalling $t = 1/s$, and

Step 3: Applying the inverse Laplace transform to the $\left[\frac{3}{3}\right]$ approximant, we then obtain the approximate solution of the problem.

$$y(t) = a_1 e^{-0.0894697t} + a_2 e^{(-0.259986-4.97021i)t}((0.993673+0.112314i) + a_3 e^{(0.1+9.94041i)t}) \tag{4.26}$$

where

$a_1 = 0.0182781, a_2 = (0.0408609-0.00230191i), a_3 = (1+0.1i)$

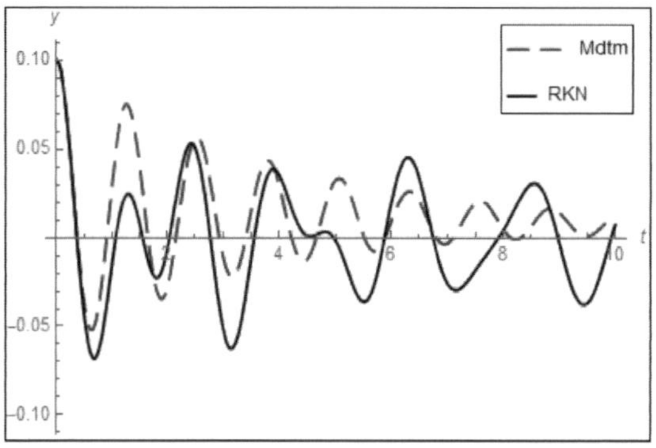

Figure 6. Results from the MDTM and RKN for $\gamma = 0.5$, $\omega = 3$.

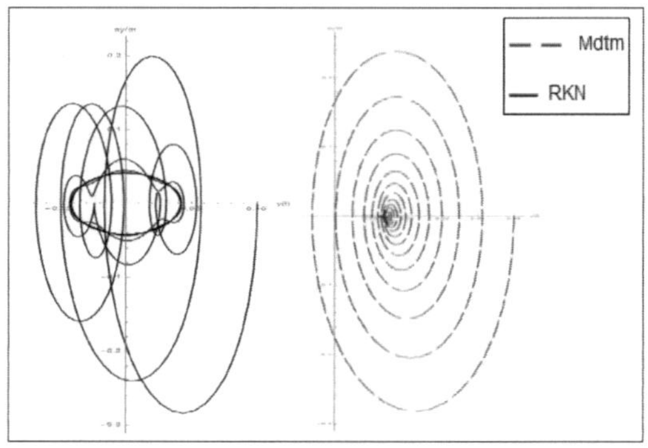

Figure 7. Phase plane diagram from the MDTM and RKN for $\gamma = 0.5$, $\omega = 3$.

Figure 6 shows the comparison results obtained by the MDTM as shown in Eq. (4.26) and RKN. Noticed that both results agreed well only for a very short period of time (i.e., t < 1 s). Clearly, increasing a frequency of the external force causes more errors.
Figure 7 shows that the result is in agreement with a phase plane diagram that only demonstrate the stable equilibrium point for MDTM in case of larger frequency of the external force.

5. Conclusion

The results from the MDTM and RKN methods are close to each another when the amplitude and frequency of an external force are small. However, the results are quite different when the amplitude and frequency of an external force are larger. From the phase plane diagram, the MDTM shows the stable equilibrium point. Therefore the MDTM is considered the suitable method to find approximate solutions of the non-linear Duffing oscillator with an external force.

6. References

[1] Jang MJ, Chen CL, Liy YC. On solving the initial-value problems using the differential transformation method. Appl Math Comput 2000;115: 145–60.
[2] Köksal M, Herdem S. Analysis of nonlinear circuits by using differential Taylor transform. Comput Electr Eng 2002;28:513–25
[3] S. Momani. and V. S. Erturk. 2008. "Solutions of non-linear oscillator by the modified differential transform method." Computer and Mathematics with Application 55. : 833-842.
[4] Tonya De George. and Anne Marie Marshall. 2009. "The Duffing equation: A nonlinear differential equation."
[5] Abolfazl Jalivand. And Hadi Fotoohabadi. 2011. "The application of Duffing oscillator in weak signal detection." Electronics and Communications. 9(1):1-6
[6] G.A. Baker. 1975. Essentials of Pade approximants. London : Academic Press.

IOP Conf. Series: Journal of Physics: Conf. Series **814** (2017) 012020 doi:10.1088/1742-6596/814/1/012020

Application of exponential homotopy algorithm in the inverter harmonic elimination

Li Zhang[1,2,3]

[1] Data Recovery Key Laboratory of Sichuan Province, Neijiang Normal University, Neijiang 641110, China
[2] Numerical Simulation Key Laboratory of Sichuan Province, Neijiang Normal University, Neijiang 641110, China
[3] College of Mathematics and Information Science, Neijiang Normal University, Neijiang 641110, China

e-mail: lizhang0212@126.com

Abstract. Eliminating harmonic pollution and improving power factor are the quite important task in the field of power electronics. From the mathematical point of view, harmonic elimination problems can be translated into nonlinear equations. But it is difficult to directly solve the nonlinear equations because of complexity. For this reason, exponential homotopy method is proposed based on homotopy method in this paper. It has focused on built up homotopy equations by modifying the singularities of Jacobian matrix, and on this basis homotopy equations are transformed into the differential initial value problems. Numerical results show that the new exponential homotopy method has higher precision than other algorithms, and the singularity is improved.

1. Introduction

In recent years, special harmonic elimination technology [1] (SHET) is proposed by Patel H.S and Hoft R.G in 1973, which is a pulse width modulation technology and has been widely used in frequency control devices and other projects [2]. Due to its many advantages, such as high quality output waveform, lower switching frequency, weak loss, etc., it has been a hot concern [3-7]. Since trans-cenddental nonlinear equations often are involved in the harmonic elimination model with more dependence on the initial, iterative methods have poor convergence, even divergent for solving. Based on these reasons, many experts and scholars achieved good results by using homotopy algorithms [8-10]. Despite homotopy algorithm is a very effective method, it is difficult to avoid singularity [11-13]. Here, we propose the exponential homotopy method, which is continuous with large-scale convergence, and Jacobi matrix of new function is nonsingular. Furthermore, the solution of new function is in accordance with equations, and has better effect. Exponential homotopy algorithm not only improves the convergence speed and accuracy, but also has a small amount of calculation. It is a method for solving harmonic elimination equations, which close enough to the real solution.
Based on this observation in this paper, we give exponential homotopy method in the next section. In section 3, PWM model has been stated and discussed. Some numerical experiments show feasibility and validity of it.

2. Exponential homotopy method

Let D be a domain in $\mathbf{R}^n$, we consider following nonlinear equations:

$$\boldsymbol{F}(\boldsymbol{x})=\mathbf{0} \tag{1}$$

where $\boldsymbol{F}(\boldsymbol{x})=\left(f_1(\boldsymbol{x}),f_2(\boldsymbol{x}),\cdots,f_n(\boldsymbol{x})\right)^T$, $\boldsymbol{x}=(x_1,x_2,\cdots x_n)^T$. It is well known that nonlinear mappings $f_i(\boldsymbol{x})$ $(i=1,2,\cdots)$ are defined in domain D.
We focus on the following form:

$$\boldsymbol{H}(\boldsymbol{x},s)=\boldsymbol{F}(\boldsymbol{x})+(s-1)\boldsymbol{P}(\boldsymbol{x})=\mathbf{0} \tag{2}$$

with $s\in[0,1]$, $\boldsymbol{x}\in D$. The solution of equation (2) is $\boldsymbol{x}_0$ as $s=0$, and when $s=1$, satisfying $\boldsymbol{H}(\boldsymbol{x},1)=\boldsymbol{F}(\boldsymbol{x})=\mathbf{0}$ .Taking $\boldsymbol{P}(\boldsymbol{x})=\boldsymbol{G}(\boldsymbol{x})\boldsymbol{F}(\boldsymbol{x}_0)$, where $\boldsymbol{G}(\boldsymbol{x})=\underset{i}{diag}(e^{-ux_i})$ is diagonal matrix, which e^{-ux_i} $\left(i=1,2,\cdots,n\right)$ are the main diagonal elements. u is a parameter. Obviously, for any $\boldsymbol{x}\in\mathbf{R}^n$ is nonsingular. Thus homotopy equation (2) can be rewritten as

$$\boldsymbol{F}(\boldsymbol{x})+(s-1)\underset{i}{diag}(e^{-ux_i})\boldsymbol{F}(\boldsymbol{x}_0)=\mathbf{0} \tag{3}$$

Let $s=1-e^{-t}$, we have

$$\boldsymbol{F}(\boldsymbol{x})-e^{-t}\underset{i}{diag}(e^{-ux_i})\boldsymbol{F}(\boldsymbol{x}_0)=\mathbf{0} \tag{4}$$

with $t\in[0,+\infty)$, $\boldsymbol{x}\in D$. Here, Eq. (4) is called exponential homotopy method.

Equation (4) is equivalent with

$$e^{-t}\boldsymbol{P}(\boldsymbol{x})=e^{-t}\underset{i}{diag}(e^{-ux_i})\boldsymbol{F}(\boldsymbol{x}_0) \tag{5}$$

According to the definition of function Jacobi matrix, we derive that $P'(\boldsymbol{x})=-u\,\underset{i}{diag}(e^{-ux_i})\,\underset{i}{diag}\,f_i(\boldsymbol{x}_0)$.
Here derivation to equation (5) at t, we get

$$\boldsymbol{F}'(\boldsymbol{x})\boldsymbol{x}_t'-e^{-t}\underset{i}{diag}(e^{-ux_i})[-\boldsymbol{F}(\boldsymbol{x}_0)-u\,\underset{i}{diag}\,f_i(\boldsymbol{x}_0)\boldsymbol{x}_t']=0 \tag{6}$$

Combining (6) with relation (5), we obtain

$$\boldsymbol{F}'(\boldsymbol{x})\boldsymbol{x}_t'+\boldsymbol{F}(\boldsymbol{x})+u\,\underset{i}{diag}(f_i(\boldsymbol{x}))\boldsymbol{x}_t'=0 \tag{7}$$

Furthermore,

$$\boldsymbol{x}_t'=-[\boldsymbol{F}'(\boldsymbol{x})+u\,\underset{i}{diag}(f_i(\boldsymbol{x}))]^{-1}\boldsymbol{F}(\boldsymbol{x}) \tag{8}$$

Therefore, we get the initial value problem of ordinary differential equation

$$\begin{cases}\dfrac{\mathrm{d}\boldsymbol{x}}{\mathrm{d}t}=-[\boldsymbol{F}'(\boldsymbol{x})+u\,\underset{i}{diag}\,f_i(\boldsymbol{x})]^{-1}\boldsymbol{F}(\boldsymbol{x}),\\ \boldsymbol{x}(0)=\boldsymbol{x}_0,\qquad t\in[0,+\infty).\end{cases} \tag{9}$$

Generally speaking, when $\boldsymbol{F}(\boldsymbol{x})$ is singular or approach singularity, u can be adjusted. There exists u such that $\boldsymbol{F}'(\boldsymbol{x})+u\,\underset{i}{diag}\,f_i(\boldsymbol{x})$ is nonsingular, as long as u is large enough. Here, we adopt midpoint quadrature formula [14] to solve equation (9), then combining with Newton method, we get the iterative schemes:

$$\begin{cases} \boldsymbol{x}_1 = \boldsymbol{x}_0 - h\left[G(\boldsymbol{x}_0)\right]^{-1} \boldsymbol{F}(\boldsymbol{x}_0), \\ \boldsymbol{x}_{k+\frac{1}{2}} = \boldsymbol{x}_k - \dfrac{1}{2}(\boldsymbol{x}_k - \boldsymbol{x}_{k-1}), \qquad k = 1, 2, \cdots, N-1 \\ \boldsymbol{x}_{k+1} = \boldsymbol{x}_k - h[G(\boldsymbol{x}_{k+\frac{1}{2}})]^{-1} \boldsymbol{F}(\boldsymbol{x}_{k+\frac{1}{2}}). \end{cases} \tag{10}$$

and

$$\boldsymbol{x}_{k+1} = \boldsymbol{x}_k - \boldsymbol{F}'(\boldsymbol{x}_k)^{-1} \boldsymbol{F}(\boldsymbol{x}_k), \quad k = N, N+1, \cdots \tag{11}$$

where $G(\boldsymbol{x}) = \boldsymbol{F}'(\boldsymbol{x}) + u\, \underset{i}{diag}\, f_i((\boldsymbol{x}))$, h is a step.

3. Harmonic elimination model

Taking the single phase SHEPEM as example, figure 1 is known to be unipolar PWM waveform of switching.

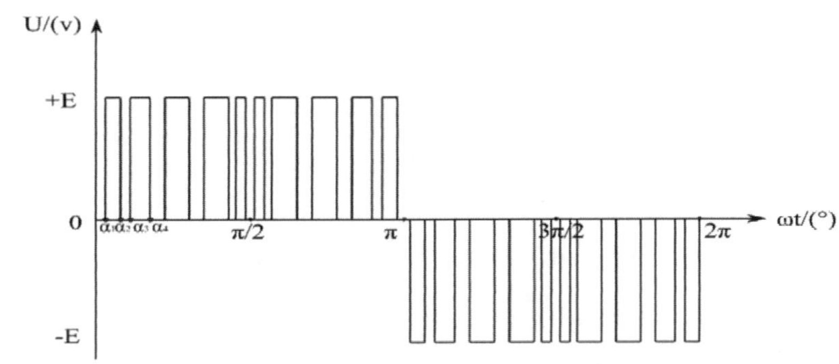

Figure 1. Unipolar control pulse waveform

Fourier expansions of its output voltage are expressed as

$$f(\omega t) = \sum_{n=1}^{\infty} [a_n \sin(n\omega t) + b_n \cos(n\omega t)] \tag{12}$$

where

$$\begin{cases} a_n = \dfrac{1}{\pi} \displaystyle\int_0^{2\pi} f(\omega t) \sin(n\omega t) \mathrm{d}(\omega t), \\ b_n = \dfrac{1}{\pi} \displaystyle\int_0^{2\pi} f(\omega t) \cos(n\omega t) \mathrm{d}(\omega t). \end{cases} \tag{13}$$

The output voltage waveform in figure 1 is both odd functions and odd harmonics. That is to say, for $f(\omega t)$, $\dfrac{\pi}{2}$ is the axis of symmetry in $[0, \pi]$, and π is symmetry point in $[0, 2\pi]$. Then we have

$$\begin{cases} f(\omega t) = -f(\omega t + \pi), \\ f(\omega t) = f(\omega t - \pi). \end{cases} \tag{14}$$

Eq. (14) is substituted in relation (13), we get

$$\begin{cases} b_n = 0, & n = 1,2,3,\cdots \\ a_n = 0, & n = 2,4,6,\cdots \\ a_n = -\dfrac{4E}{n\pi}[1 + 2\sum_{k=1}^{N}(-1)^k \cos(n\alpha_k)] & n = 1,3,5,\cdots \end{cases} \tag{15}$$

where E is voltage of direct-current generatrix, N is the number of switching, α_k is the k-th switching in N and n is frequency of harmonics.
The amplitude of the fundamental is

$$a_1 = -\frac{4E}{\pi}[1 + 2\sum_{k=1}^{N}(-1)^k \cos\alpha_k] \tag{16}$$

Let U_m be the selected amplitude of the fundamental, and the other amplitudes of higher harmonics be zero, PWM model is described by

$$\begin{cases} a_1 = U_m, \\ a_n = 0. & n = 3,5,\cdots \end{cases} \tag{17}$$

To solve relation (17) with N dimensions nonlinear equations, we arrive at a group switch angles in $[0, \pi/2]$ and satisfy the conditions:

$$0 < \alpha_1 < \alpha_2 < \alpha_3 < \alpha_4 < \alpha_5 < \alpha_6 < \alpha_7 < \alpha_8 < \cdots < \pi/2 \tag{17}$$

4. Numerical examples

In order to further illustrate the validity of exponential homotopy method in harmonic elimination equations, Newton method, midpoint quadrature formula Newton homotopy method (MQFNM) are compared with exponential homotopy method by C++ software.

4.1 Numerical example 1

The parameters of PWM model are shown in Table 1.

Table 1. The parameters of numerical example

Number of switch angle (N)	Initial value(α_0)	error precision(ε)
8	$[5, 20, 30, 40, 45, 60, 65, 89]$	10^{-6}

The results of iteration solution and final exact solution are compared in Table 2, and the number of iteration is shown in Table 3.

Table 2. Comparison of iteration solution by three methods

α	α_0	Newton method	MQFNM solution	Exponential homotopy solution	Exact solution
α_1	5	8.970	8.681	8.612	8.689
α_2	20	20.156	20.746	20.133	20.023
α_3	30	26.425	26.440	26.505	26.563
α_4	40	40.543	40.327	40.491	40.421
α_5	45	44.694	44.923	44.764	44.725
α_6	60	63.346	63.774	63.243	63.214
α_7	65	64.337	64.906	64.592	64.578

α_8	89	89.943	89.725	89.881	89.884

Table 3. Comparison of iteration number by three methods

Method	Newton method	MQFNM solution	Exponential homotopy solution
Number of iteration	4	4	4

As can be seen from Table 2 and Table 3, it is clear that three methods converge fast to the accurate solution in a few iterations. The reason was that initial values are very close to the accurate values.

4.2 Numerical example 2

The parameters of PWM model are shown in Table 4.

Table 4. The parameters of numerical example

Number of switch angle (N)	Initial value(α_0)	error precision(ε)
8	$[5, 20, 25, 40, 45, 60, 65, 84.452]$	10^{-6}

The results of iteration solution and final exact solution are compared in Table 5, and the number of iteration is shown in Table 6.

Table 5. Comparison of iteration solution by three methods

α	α_0	Newton method	MQFNM solution	Exponential homotopy solution	Exact solution
α_1	5	9.970	9.681	8.444	8.472
α_2	20	22.157	22.744	21.123	21.157
α_3	25	38.024	38.140	37.105	37.160
α_4	40	43.551	43.238	42.193	42.249
α_5	45	56.628	56.923	56.176	56.259
α_6	60	divergence	63.798	63.124	63.207
α_7	65	divergence	75.906	75.794	75.905
α_8	84.452	83.943	83.881	83.770	83.881

Table 6. Comparison of iteration number by three methods

Method	Newton method	MQFNM solution	Exponential homotopy solution
Number of iteration	13	10	5

As can be seen from Table 2 and Table 3, we find that there may be divergence phenomena in Newton method, and exponential homotopy method is much better regarding iterations and convergence than Newton method, midpoint quadrature formula-Newton homotopy method. Therefore, results show that exponential homotopy method not only has been improved significantly for efficiency, but also convergence effects are very good.

5. Conclusion

Through depth studying the inverter harmonic elimination model, we propose an effective method to solve nonlinear equations in model. Of course, there is also a bipolar pulse model, the principle exactly as same as that of single pulse model. Through simulation experiments, the results of specific examples show its efficiency and improving the accuracy of solution. In general, it has good results to solve practical problems of harmonic elimination, and provides a possibility for realization of real-time harmonic elimination. If the algorithm can be better, it will be a major breakthrough in harmonic elimination field.

6. References

[1] Patel HS; and Hoft RS P 1973 *Generalized Technique of Harmonic Elimination and Voltage Control in the Thyristor Inverter: Part I-Hamonic Elimiation* vol 3 (IEEE Trans. on Industry Application) p 301-317
[2] Li XX, et al 1996 *Pulse Width Modulation* (Wuhan: Huazhong University of Technology Press)
[3] Tong WM Chen XY and Hu Y et al P 1998 *Investigation of Solution to the Nonlinear Equation Set up of SHET in Converters* vol 5 (Proceedings of the Csee) p 357-360
[4] Fei WM, Lv ZY and Yao WX P 2003 *Research on Selected Harmonic Elimination PWM Technique Applicable to Three-level Voltage Inverters* vol 9 (Proceedings of the Csee) p 11-15.
[5] Huang YQ and Xie YX P 2003 *Algorithm Research of the Inverse Matrix in the Solution of the Inverter Harmonic Elimination Model* vol 11 (Journal of South China University of Technology (Natural Science)) P 1-4
[6] Zhang YL Fei WM and Lv ZM et al P 2004 *Research of Selected Harmonic Elimination PWM Technique Applicable to Three-level Voltage Inverters* vol 19(Transactions of China Electrotechnical Society) p 16-20
[7] Zhao YC and Zhao ZM P 2007 *Multiple Solution for Selective Harmonic Eliminated PWM Applied to Three-Level Inverter* vol 22 (Transactions of China Electrotechnical society) p 74-78
[8] Li ZD, Zhou QY and Li H et al P 2004 *A Novel Algorithm for Real-time Solution of Nonlinear Surmount SHET Equations* vol 22 (Journal of Northwestern Polytechnical University) p 37-39
[9] Li X Liu JP and Yang AM P 2007 *Study and Application of Homotopy New Algorithm in Inverter Harmonic Elimination Mod*el vol 25 (Journal of Jiamusi University(Natural Science Edition),) p 381-384
[10] Zhu MJ Zhao P and Peng H et al P 2004 *Fast Homotopy Algorithm of the Inverter Harmonic Elimination PWM Model* vol 2(Journal of South China University of Technology (Natural Science)) p 28-30
[11] Huang XD, Zeng ZG and Ma YN 2004 *The Theory and Methods for Nonlinear Numerical Analysis*(Wuhan: Wuhan University Press)
[12] Li QY P 1980 *Solution of Systems of Nonlinear Equations by Differential Continuation* vol 1(Journal on Numerical Methods and Computer Applications) p45-52
[13] Li SB P 1983 *Function Factor Method-A New Method of Passing the Singularities Which Arise in the Continuation Methods for Solving Systems of Nonlinear Equations* vol 2 (Mathematica Numerica Sinica) p162-175
[14] Li QY Mo ZZ and Qi LQ 1987 *Numerical Methods of Nonlinear Equations*（Beijing: Science Press）

Acknowledgments

This work was financially supported by the National Natural Science Foundation of China (11502121), the education department of Sichuan Province university innovation team (13TD0001), the education department of Sichuan Province project (16ZA0310), and the major achievements transformation project of Neijiang Normal University (14CZ02).

IOP Conf. Series: Journal of Physics: Conf. Series **814** (2017) 012021 doi:10.1088/1742-6596/814/1/012021

Non-orthogonal frequency shift keying for optical wireless link based on visible light emitting diode

Yong-Yuk Won[1], **Sang Min Yoon**[2], **and Dongsun Seo**[1]

[1]Dept. of Electronic Engineering, Myongji University, 116 Myongji-ro, Cheoin-gu, Yongin, Gyeonggi-do, 17058, South Korea
[2]School of Computer Science, Kookmin University, 77 Jeongneung-ro, Seongbuk-gu, Seoul, 02707, South Korea

bluejerry@mju.ac.kr

Abstract. A non-orthogonal frequency shift keying (NOFSK) modulation technique, which improves the transmission capacity of an optical wireless link based on a visible light emitting diode (LED), is proposed in this paper. A NOFSK signal with the modulation order of 4 is generated and then is transmitted using an optical wireless link based on the white LED in order to verify experimentally the proposed technique. One more bit per symbol are transmitted successfully showing the only 3% increase of 2-MHz physical bandwidth at the bit error rate (BER) below 10^{-5}.

1. Introduction

In recent years, a visible light emitting diode (LED) as a solid-state semiconductor device has been playing a major role in illumination instead of a fluorescent lamp and an incandescent one [1]. Furthermore, it has been verified experimentally that the visible LED can be utilized as an optical wireless transmitter in a visible light communication (VLC) system [2-5]. However, the frequency response of the visible LED is usually limited about 5 MHz due to the relatively slow relaxation time of the phosphor [2].

In this conference paper, a non-orthogonal frequency shift keying (NOFSK) is proposed in order to improve the transmission capacity of an optical wireless link based on the visible LED with frequency response below 5 MHz. It is well known that the existing orthogonal frequency shift keying (FSK) shows a relatively low bandwidth efficiency because the frequency spacing between adjacent radio frequency (RF) carriers should be determined considering their occupied channel bandwidth to avoid the inter carrier interference (ICI) effect [6]. However, our proposed NOFSK technique allows us to have a much smaller frequency spacing than the channel bandwidth because a received NOFSK signal is correlated with only the single RF carrier at a time in order to recover the NOFSK symbols. Besides, there is no flicker effect in the VLC system because the intensity of RF carrier is not changed during the modulation of NOFSK signal.

We implement experimentally an optical wireless link based on the visible LED modulated by a NOFSK signal with the modulation order of 4 in the VLC link in order to demonstrate that the proposed technique has two times more channel capacity than on-off keying (OOK) within 5-MHz physical bandwidth. For your information, it means that the only OOK data below 5 Mbit/s can be transmitted using the visible LED within 5-MHz physical bandwidth. Therefore, it is important to check if the transmission rate of 10 Mbit/s is established in case of using the proposed 4-NOFSK scheme or not.

IOP Conf. Series: Journal of Physics: Conf. Series **814** (2017) 012021 doi:10.1088/1742-6596/814/1/012021

2. NOFSK modulation & demodulation

Figure 1 shows how multi-level(m) NOFSK modulated signal is generated and then the m-NOFSK symbols are recovered. Figure 1(a) shows the modulation scheme of m-NOFSK signal, while figure 1(b) shows the demodulation of m-NOFSK signal. As shown in figure 1(a), all RF carriers (f1, f2, ..., fm) with sampling points of n are sampled digitally and then are generated using a software switch based on digital signal processing (DSP). The m-NOFSK symbols, which are encoded with log2m bits, is allocated to one of all RF carriers. The m-NOFSK symbols are given by Eq. (1).

$$x_s = A_s \sum_{k=1}^{m} \sum_{i=1}^{n} \cos\{2\pi(f_c + k\Delta f)t(i)\} \quad (1)$$

where A_s is the energy per symbol, f_k is the frequency of digitally sampled RF carrier, f_c is the angular frequency of the RF carrier, Δf is the frequency separation between adjacent RF carriers, and t(i) is a discrete sampled time from t(1) to t(n).

Figure 1(b) shows the block diagram of the m-NOFSK signal demodulation. A received m-NOFSK signal is correlated sequentially with only one local RF carrier at a time. For example, in the 1st step of DSP based correlation, other local RF carriers were disconnected while a local RF carrier (f1) is correlated with the received m-NOFSK signal during the duration of m-NOFSK one symbol. The DSP based correlation is expressed as shown in Eq. (2).

$$Cor(f_{j,k}) = \frac{\sum_{j=1}^{m} \sum_{k=1}^{m} \sum_{i=1}^{n} B_s \cos\{2\pi(f_c+k\Delta f)t(i)+\phi(k)\}\cos(2\pi f_j t(i))}{\sum_{k=1}^{m} \sum_{i=1}^{n} B_s \cos^2\{2\pi(f_c+k\Delta f)t(i)+\phi(k)\}} \quad (2)$$

where f_j is the frequency of local RF carrier, B_s is the intensity of received m-NOFSK signal, and $\phi(k)$ is the phase change of the received m-NOFSK signal. As shown in figure 1(b), after correlations between the received m-NOFSK signal and all local RF carriers are executed repeatedly, all correlation coefficients (r_1, r_2, ..., r_m) are obtained and then are stored in sequential order. Next, the largest correlation coefficient (r_{max}) is chosen among the stored ones and then m-NOFSK symbol, which is mapped to r_{max}, is recovered in the demodulation part. Therefore, there is no ICI effects in the proposed technique because the received m-NOFSK signal is not correlated with all local RF carriers at the same time, unlike the existing FSK scheme. Furthermore, it is possible to have the much smaller frequency spacing between adjacent RF carriers than the channel bandwidth of m-NOFSK signal because of the sequential and repetitive correlation based on DSP.

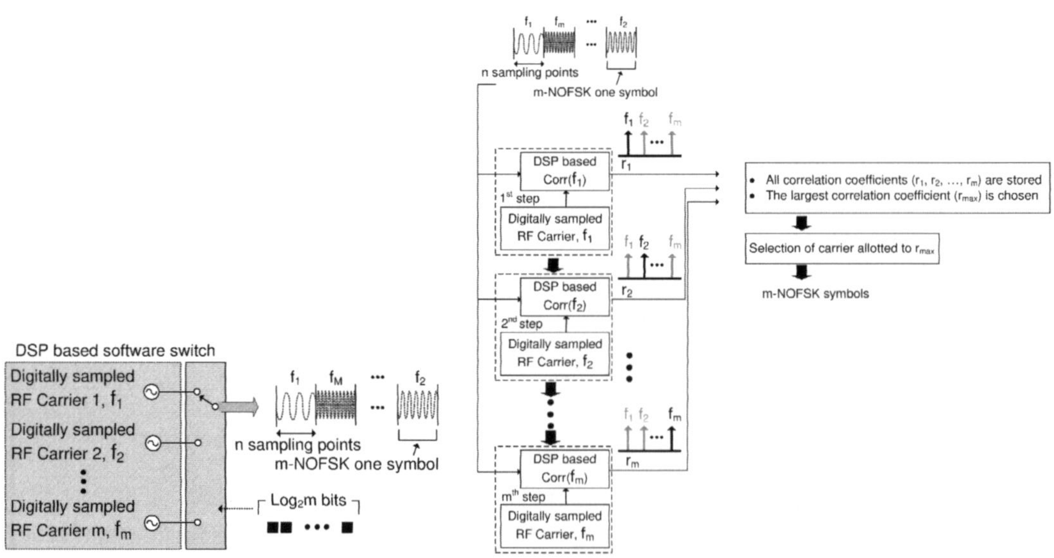

(a) (b)

Figure 1. Non-orthogonal frequency shift keying scheme, (a) modulation blocks (b) demodulation blocks.

3. Experimental setup

Figure 2 shows the experimental setup of which we check if the transmission capacity of the optical wireless link based on the visible LED is improved by using the proposed technique or not. An inset of figure 2 was the channel response of optical wireless link shown in figure 2. The 3-dB bandwidth below 2.5 MHz was observed as shown in the inset of figure 2. The modulation order of NOFSK signal was set to 4. A 4-NOFSK symbol was encoded using 2^{15}-1 pseudo-random binary sequence in MATLAB®. All RF carriers (2 MHz, 2.02 MHz, 2.04 MHz, and 2.06 MHz) were sampled digitally using DSP to generate and recover the 4-NOFSK symbols. The frequency spacing was 1% of 2-MHz physical bandwidth. This indicates that the proposed technique was implemented based on non-orthogonality without considering the channel bandwidth of 4-NOFSK signal.

The 4-NOFSK signal, which is generated by MATLAB®, is converted into an electrical one with the help of an arbitrary waveform generator (AWG: Tektronix 70002A). The sampling rate of the AWG was 250 Msamples/s. The number of sampling point per 4-NOFSK one symbol was 125. The 4-NOFSK signal waveform was combined with a dc source (bias current: 300 mA) using bias-T and then the diffused light of the visible LED was modulated directly. The modulated light was transmitted wirelessly to Avalanche photo diode (APD) with a frequency response of 100 MHz. It was located one meter away from the visible LED. A biconvex glass lens (Thorlab, LB1723, focal length: 60 mm) was used to increase the light collecting efficiency. The illumination of 100 lux was measured in front of the APD. The waveform of the received 4-NOFSK signal was recorded using an oscilloscope (Tektronix MSO 71604C) at a sampling rate of 1 Gsamples/s. Next, 4-NOFSK signal waveform was correlated with local RF carriers of the same frequency as 4-NOFSK modulation using the DSP based correlation of Eq. (2) to recover 4-NOFSK symbols.

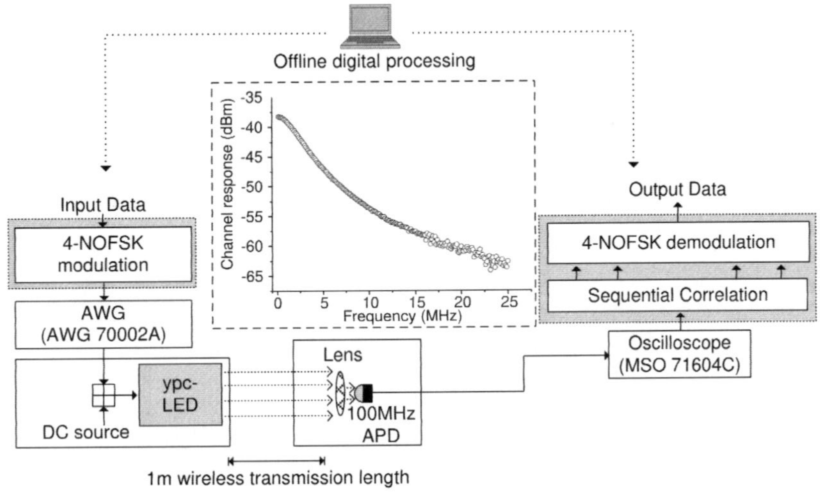

Figure 2. Experimental wireless 4-NOFSK link based on visible LED, center inset: measured frequency response of ypc-LED.

4. Experimental results

At first, it is necessary to check that the channel capacity of optical wireless link using the visible LED is increased using the proposed NOFSK technique. For this, we measured the RF spectra of 4-NOFSK and OOK signals.

Figure 3 shows the RF spectra of 4-NOFSK and OOK signals at the input signal to noise ratio (SNR) of 15 dB. As shown in figure 3, the full width half maximum (FWHM) bandwidth of 4-NOFSK signal was increased a little larger than that of OOK signal by 60 kHz. This result means that the proposed technique can allow us to transmit one more bits per symbol with an increase of only 3% from the bandwidth (2 MHz) of OOK. Therefore, it was implied that the bandwidth efficiency of visible LED was improved at double that without the proposed technique.

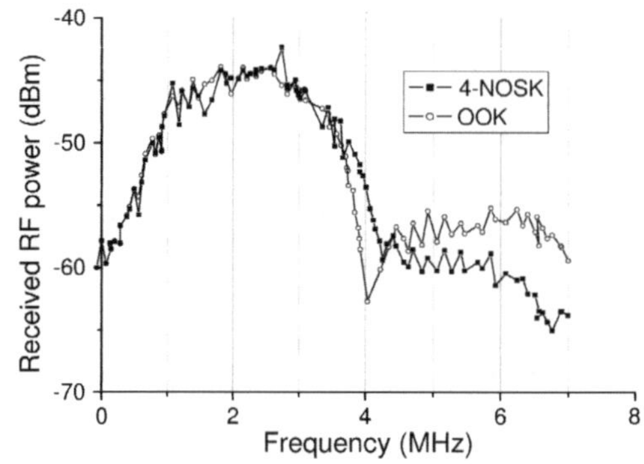

Figure 3. RF spectra of 4-NOFSK and OOK signals

Figure 4 shows the measured BER curves of two kinds of signal (4-NOFSK, and OOK) as the input SNR increases. The 2-dB SNR penalty was observed at the BER of 10-3 in comparison with OOK signal. Also, we need to check the non-orthogonality of the proposed technique. This can be verified by investigating the impact of ICI effects among RF carriers for 4-NOFSK scheme as their frequency spacing get narrower. This results will be presented in this conference.

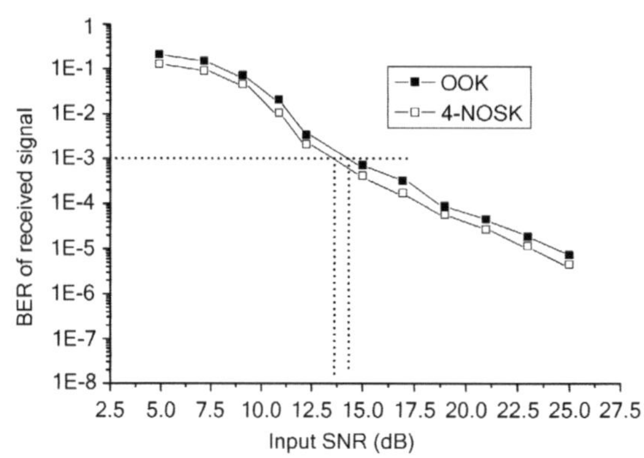

Figure 4. Measured BER curves against the input signal to noise ratio.

5. References

[1] D. Karunatilaka, F. Zafar, V. Kalavally, 2015 *IEEE Commun. Surv. & Tutor.* **17**, 3
[2] J. Gruber, S. C. J. Lee, K.-D. Langer, T. Koonen, J. W. Walewski, 2007 Proc. Eur. Conf. Optical *Communication (Berlin) (IEEE/OSA)* p 1
[3] J. Vucic, C. Kottke, S. Nerreter, K.-D. Langer, J. W. Walewski, 2010 *J. Lightw. Technol.* **28**, 24
[4] C. Kottke, J. Hilt, K. Habel, J. Vucic, K.-D. Langer, 2012 *Proc. Eur. Conf. Optical Communication (Amsterdam) (IEEE/OSA)* p We.3.B.4
[5] P.A. Haigh, Z. Ghassemlooy, S. Rajbhandari, I. Papakonstantinou, W. Popoola, 2014 *J. Lightw. Technol.* **32**, 9
[6] B. Sklar, 2001 *Digital Communications: Fundamentals and Applications*, (Prentice Hall)

Acknowledgments
This research was supported by Basic Science Research Program through the National Research Foundation of Korea(NRF) funded by the Ministry of Education(NRF-2016R1D1A1B01008748).
Sang Min Yoon was also supported by the IITP grant funded by the Korean government (MSIP, No. R0115-16-1009: An Interaction Platform for Learning Foreign Languages with Tangible Reality Based on Experiential Awareness).

Improved synchronization criteria for time-delayed chaotic Lur'e systems using sampled-data control

Wenyong Duan[1],Yan Li[2], Xiaorong Fu[1], Baozhu Du[3]
[1]School of Electrical Engineering, YanCheng Institute of Technology, YanCheng JiangSu, China
[2]Dean's Office, Yancheng Biological Engineering Higher Vocational Technology School, YanCheng JiangSu, China
[3]School of Automation, Nanjing University of Science and Technology, Nanjing, Jiangsu, China

dwy1985@126.com

Abstract. This paper is concerned with the synchronization for a class of time-delayed chaotic Lur'e systems using sampled-data control. Both of time-varying and time-invariant delays are considered. New criteria are proposed in terms of linear matrix inequalities (LMIs) by employing a modified LKF combined with the delay-fraction theory and some novel terms. The criteria are less conservative than some previous ones and a longer sampling period is achieved under the new results. Furthermore, the derived conditions are employed to design a sampled-data controller. The desired controller gain matrix can be obtained by means of the LMI approach. Finally, a numerical examples and simulations on Chua's circuit is presented to show the effectiveness of the proposed approach.

1. Introduction

The synchronization of chaotic system has arisen a very hot topic in the nonlinearity community since the pioneering work of [1], where the chaotic system synchronization was first studied. Meanwhile, time-delays are commonly encountered in practice and are often attributed as a source of poor performance and instability. Thus, the synchronization of chaotic Lur'e systems with time delays attains considerable significance. Note that recent improvements of the synchronization criteria for chaotic Lur'e systems with time-invariant delays have been obtained in [2-8]. In order to further reduce the conservativeness of the synchronization criteria in [7,8], a novel integral inequality was developed based on the celebrated Jensen's integral inequality [9] and Wirtinger integral inequality [10]. However, there still exists room for improving the synchronization criteria of the above literature.

Notation: The notation $P>0$ (<0) mean that matrix P is positive (negative) definite. I denotes an identity matrix with appropriate dimensions. $*$ denotes the symmetric terms in a block matrix and $\mathrm{diag}\{\cdots\}$ denotes a block-diagonal matrix. e_i ($i=1,\cdots,m$) are block entry matrices. For example, $e_2^T=\mathrm{diag}\{0,I,0,\cdots,0\}$. $Sym\{B\}=B+B^T$.

2. Problem formulation

Consider the following general master-slave type of time-delayed chaotic Lur'e systems with sampled-data feedback controller:

$$
\begin{aligned}
&M:\begin{cases}\dot{x}(t)=Ax(t)+Bx(t-d(t))+H\sigma(D^{T}x(t)),\\ p(t)=Cx(t),\end{cases}\\
&S:\begin{cases}\dot{z}(t)=Az(t)+Bz(t-d(t))+H\sigma(D^{T}x(t))+u(t),\\ q(t)=Cz(t),\end{cases}\\
&C:u(t)=K(p(t_k)-q(t_k)),\qquad t_k\le t<t_{k+1}
\end{aligned}
\tag{1}
$$

which consists of the master system M, the slave system S and the sampled-data feedback controller C. The master system and the slaver system are identical Lur'e chaotic systems when $u(t)=0$ with state vectors $x(t)$, $z(t)\in R^{n}$, outputs of subsystems $p(t)$, $q(t)\in R^{l}$. $A\in R^{n\times n}$, $B\in R^{n\times n}$, $C\in R^{l\times n}$, $D\in R^{n_h\times n}$ and $H\in R^{n\times n_h}$. $u(t)\in R^{n}$ is the slaver system control input and $K\in R^{n\times l}$ is the sampled-data feedback control gain matrix to be designed. $d(t)$ is the time-varying delay satisfying the following condition with some scalars d_1, d_2, d_μ: $d_1\le d(t)\le d_2$, $\dot{d}(t)\le d_\mu<1$. $\sigma(\cdot):R^{n_h}\mapsto R^{n_h}$ is a diagonal nonlinearity with $\sigma_i(\cdot)$ belong to $[\omega_i^{-},\omega_i^{+}]$, i.e., $[\sigma_i(\xi)-\omega_i^{+}\xi][\sigma_i(\xi)-\omega_i^{-}\xi]\le 0$. Suppose that the updating signal (successfully transmitted signal from the sampler to the controller and to the ZOH) at the instant t_k has experienced a constant signal transmission delay η. Therefore, we have $t_{k+1}-t_k+\eta\le h+\eta=\tau_M$.

Now, we define the synchronization error as $e(t)=x-z$, $\tau(t)=t-t_k+\eta$, $h_\tau(t)=h+\eta-\tau(t)$, then $\dot{e}=Ae+Be(t-d(t))+Hf(D^{T}e)-KCe(t-\tau(t))$, where

$$[f_i(d_i^{T}e,z)-\omega_i^{+}d_i^{T}e][f_i(d_i^{T}e,z)-\omega_i^{-}d_i^{T}e]\le 0, \tag{2}$$

d_i^{T} denotes the i-*th* row of D^{T}. $K_1=diag\{\omega_1^{+},\cdots,\omega_{n_h}^{+}\}$ and $K_2=diag\{\omega_1^{-},\cdots,\omega_{n_h}^{-}\}$.

Definition 1. The master system and slave system are said to be asymptotically synchronized if and only if the error dynamical systems is absolutely stable in the sector $[K_2,K_1]$ for the equilibrium point $e(t)\equiv 0$. That is, $e(t)\to 0$ as $t\to\infty$.

Lemma 1. See Lemma 2.1 [11] for detail.

Lemma 2. See Lemma 2.2 [11] for detail.

3. Main results

$\tau_C=\eta+\frac{h}{2},\tau_V=\frac{h}{2},h_\tau(t)=\tau_M-\tau(t),d_{12}=d_2-d_1,\varepsilon^{T}=[e^{T}e^{T}(t-\tau(t))]$, $\delta^{T}=[\varsigma^{T}e^{T}(t-\tau(t))]$,

$\varsigma^{T}=\left[e^{T}\int_{t-\eta}^{t}e^{T}(s)ds\int_{t-\tau_C}^{t-\eta}e^{T}ds\int_{t-\tau_M}^{t-\tau_C}e^{T}ds\right]$, $\xi_1^{T}=\Big[e^{T}e^{T}(t-d_1)e^{T}(t-d(t))e^{T}(t-d_2)e^{T}(t-\eta)e^{T}(t-\tau_C)$
$e^{T}(t-\tau(t))e^{T}(t-\tau_M)\frac{1}{\eta}\int_{t-\eta}^{t}e^{T}(s)ds\frac{1}{\tau(t)-\eta}\int_{t-\tau_C}^{t-\eta}e^{T}ds\frac{1}{\tau_C-\tau(t)}\int_{t-\tau_C}^{t-\tau(t)}e^{T}ds\frac{1}{\tau_V}\int_{t-\tau_M}^{t-\tau_C}e^{T}ds\dot{e}^{T}f^{T}(\cdot)\Big]$,

$\xi_2^{T}=\Big[e^{T}e^{T}(t-d_1)e^{T}(t-d(t))e^{T}(t-d_2)e^{T}(t-\eta)e^{T}(t-\tau_C)e^{T}(t-\tau(t))e^{T}(t-\tau_M)$
$\frac{1}{\eta}\int_{t-\eta}^{t}\frac{1}{\tau_V}\int_{t-\tau_C}^{t-\eta}e^{T}ds\frac{1}{\tau(t)-\tau_C}\int_{t-\tau(t)}^{t-\tau_C}e^{T}ds\frac{1}{\tau_M-\tau(t)}\int_{t-\tau_M}^{t-\tau(t)}e^{T}ds\dot{e}^{T}f^{T}(\cdot)\Big]$.

Theorem 1. For given scalars $\eta,h,d_1,d_2>0$, $d_\mu<1$, $\$K_1,K_2$ and a scalar r, the master system M and slave system S in (1) are synchronous if there exist positive matrices $P=[P_{ab}]_{4\times 4}$, Q_i, W_i, R_j, Z_j ($i=1,2,3; j=1,2$), $\overline{G}=[G_{ab}]_{2\times 2}$, $\overline{Y}=[Y_{ab}]_{2\times 2}$ and $\overline{N}=[N_{ab}]_{2\times 2}$, positive definite diagonal matrices $\Lambda=\mathrm{diag}\{\lambda_1,\cdots,\lambda_{n_h}\}$, $\Delta=\mathrm{diag}\{\delta_1,\cdots,\delta_{n_h}\}$, T_m ($m=1,2$), any appropriately dimensioned matrices $U=[U_1^T U_2^T]^T$, $\overline{M}=[M_{ab}]_{2\times 2}$, $\overline{V}=[V_{ab}]_{2\times 2}$, T,V,L,X_1,X_2 and X_3 such that the following LMIs hold for $k=1,2$:

$$\begin{bmatrix} R_2 & T \\ * & R_2 \end{bmatrix}>0,\ \begin{bmatrix} P_{11}+\tau_M Sym\{X_1\} & P_{12} & P_{13} & P_{14} & \tau_M X_3 \\ * & P_{22} & P_{23} & P_{24} & 0 \\ * & * & P_{33} & P_{34} & 0 \\ * & * & * & P_{44} & 0 \\ * & * & * & * & \tau_M Sym\{X_2\} \end{bmatrix}>0, \tag{3}$$

$$\Theta_i^k=\Pi_i^k+\tau_M Sym\{[e_1\quad e_7]\overline{X}[e_{13}\quad 0]^T\}+\tau_M[e_7\quad e_{13}]\overline{N}[e_7\quad e_{13}]^T+G(2,t)<0, \tag{4}$$

$$\Xi_i^k=\begin{bmatrix} \Pi_i^k-\tau_M e_7 N_{11} e_7^T+G(1,t) & \Omega_{12} \\ * & -\tau_M N_{22} \end{bmatrix}<0, \tag{5}$$

where $G(i,t)=[e_1\quad e_{14}]\begin{bmatrix} -2D^T K_1 T_i K_2 D & D^T(K_1+K_2)T_i \\ * & -2T_i \end{bmatrix}[e_1\quad e_{14}]^T$, $\Pi_1^k=Sym\{\Gamma_1^k P\Psi^T\}+\Theta_1+\Theta_2+J$ $+[e_1\quad e_7](\overline{U}-\overline{X})[e_1\quad e_7]^T-[e_1\quad e_2]R_1[e_1\quad e_2]^T+[e_7\quad e_5]\overline{Y}[e_7\quad e_5]^T+[e_5\quad e_6]\overline{G}[e_5\quad e_6]^T-[e_6\quad e_8]\overline{G}[e_6\quad e_8]^T-[e_2\quad e_3\quad e_4]\Sigma_{R_2}[e_2\quad e_3\quad e_4]^T-[e_1\quad e_5\quad e_9]\Sigma_{W_1}[e_1\quad e_5\quad e_9]^T-[e_6\quad e_8\quad e_{12}]\Sigma_{W_3}[e_6\quad e_8\quad e_{12}]^T$ $-Z_{W_2}\begin{bmatrix} \tilde{W}_2 & \overline{M} \\ * & \tilde{W}_2 \end{bmatrix}Z_{W_2}^T$, $\Pi_2^k=Sym\{\Gamma_2^k P\Psi^T\}+\Theta_1+\Theta_2+J+[e_1\quad e_7](\overline{U}-\overline{X})[e_1\quad e_7]^T-[e_1\quad e_2]R_1[e_1\quad e_2]^T$ $+[e_7\quad e_5]\overline{Y}[e_7\quad e_5]^T+[e_5\quad e_6]\overline{G}[e_5\quad e_6]^T-[e_6\quad e_8]\overline{G}[e_6\quad e_8]^T-[e_2\quad e_3\quad e_4]\Sigma_{R_2}[e_2\quad e_3\quad e_4]^T$ $-[e_1\quad e_5\quad e_9]\Sigma_{W_1}[e_1\quad e_5\quad e_9]^T-[e_5\quad e_6\quad e_{10}]\Sigma_{W_2}[e_5\quad e_6\quad e_{10}]^T-Z_{W_3}\begin{bmatrix} \tilde{W}_3 & \overline{V} \\ * & \tilde{W}_3 \end{bmatrix}Z_{W_3}^T$,

$\Omega_{12}=[-\tau_M U_1^T 00000-\tau_M(U_2^T+N_{12}^T)00000000]$, $\Gamma_1^1=[e_1\quad \eta e_9\quad \tau_V e_{10}\quad \tau_V e_{12}]$, $\Gamma_1^2=[e_1\quad \eta e_9\quad \tau_V e_{11}\quad \tau_V e_{12}]$, $\Gamma_2^1=[e_1\quad \eta e_9\quad \tau_V e_{10}\quad \tau_V e_{11}]$, $\Gamma_1^2=\Gamma_1^1$, $\Psi=[e_{13}\quad e_1-e_5\quad e_5-e_6\quad e_6-e_8]$, $\Theta_1=Sym\{e_1 D^T(K_1\Lambda-K_2\Delta)De_{13}^T+e_{13}D^T(\Delta-\Lambda)e_{14}^T\}$, $\Theta_2=Sym\{e_1[VAe_1^T+VBe_3^T+(A^TV^T-LC)e_7^T+(rA^TV^T-V)e_{13}^T+VHe_{14}^T]+e_3[B^TV^Te_7^T+rB^TV^Te_{13}^T]$ $-e_7[LCe_7^T+(V+rC^TV^T)e_{13}^T-VHe_{14}^T]-e_{13}[rVe_{13}^T-rVHe_{14}^T]\}$, $J=\mathrm{diag}\{Q_1+Z_1,Q_2-Q_1+Q_3,-(1-d_\mu)Q_2,-Q_3,$ $-Z_1,Z_2,0,-Z_2,0,0,0,0,d_1^2R_1+d_{12}^2R_2+\eta^2W_1+\tau_V^2(W_2+W_3),0\}$, $Z_{W_2}=[e_5-e_7\quad e_5+e_7-2e_{10}\quad e_7-e_6\quad e_7+e_6-2e_{11}]$, $Z_{W_3}=[e_6-e_7\quad e_6+e_7-2e_{11}\quad e_7-e_8\quad e_7+e_8-2e_{12}]$, $\tilde{W}_2=\mathrm{diag}\{W_2,3W_2\}$, $\tilde{W}_3=\mathrm{diag}\{W_3,3W_3\}$, $\overline{U}=\begin{bmatrix} U_1+U_1^T & -U_1+U_2^T \\ * & -U_2-U_2^T \end{bmatrix}$,

$$\Sigma_{R_2}=\begin{bmatrix} -R_2 & R_2-T & T \\ * & -2R_2+T+T^T & R_2-T \\ * & * & -R_2 \end{bmatrix},\ \Sigma_{W_j}=\begin{bmatrix} 4W_j & 2W_j & -6W_j \\ * & 4W_j & -6W_j \\ * & * & 12W_j \end{bmatrix},\ \overline{X}=\begin{bmatrix} X_1+X_1^T & X_3 \\ * & X_2+X_2^T \end{bmatrix}.$$

Moreover, the sampled-date controller gain matrix is given by $K=V^{-1}L$.

Proof: Construct an LKF candidate as $V(t)=\sum_{i=1}^{6}V_i(t)$ with

$$V_1=2\sum_{i=1}^{n_h}\int_0^{d_i^T e}[\lambda_i(\omega_i^+ s-f_i)+\delta_i(f_i-\omega_i^- s)]ds, V_2=\varsigma^T P\varsigma+h_\tau(t)\varepsilon^T\overline{X}\varepsilon,\ V_3=\int_{t-d_1}^{t}e^TQ_1eds+\int_{t-d(t)}^{t-d_1}e^TQ_2eds$$
$$+\int_{t-d_2}^{t-d_1}e^TQ_3eds+d_1\int_{-d_1}^{0}\int_{t+\theta}^{t}\dot{e}^TR_1\dot{e}dsd\theta+d_{12}\int_{-d_2}^{-d_2}\int_{t+\theta}^{t}\dot{e}^TR_2\dot{e}dsd\theta,\ V_4=\int_{t-\tau(t)}^{t}\begin{bmatrix} e(t-\tau(t)) \\ e(s) \end{bmatrix}\overline{Y}\begin{bmatrix} e(t-\tau(t)) \\ e(s) \end{bmatrix}^T ds$$
$$+\int_{t-\tau_C}^{t-\eta}\begin{bmatrix} e(s) \\ e(s-\tau_V) \end{bmatrix}\overline{G}\begin{bmatrix} e(s) \\ e(s-\tau_V) \end{bmatrix}^T ds+\int_{t-\eta}^{t}e^TZ_1eds+\int_{t-\tau_M}^{t-\tau_C}e^TZ_2eds,\ V_5=\eta\int_{-\eta}^{0}\int_{t+\theta}^{t}\dot{e}^TW_1\dot{e}dsd\theta$$

$$+\tau_V\int_{-\tau_C}^{-\eta}\int_{t+\theta}^{t}\dot{e}^T W_2\dot{e}dsd\theta+\tau_V\int_{-\tau_M}^{-\tau_C}\int_{t+\theta}^{t}\dot{e}^T W_3\dot{e}dsd\theta,\ V_6=h_\tau(t)\int_{t-\tau(t)}^{t}\begin{bmatrix}e(t-\tau(t))\\ \dot{e}(s)\end{bmatrix}^T \overline{N}\begin{bmatrix}e(t-\tau(t))\\ \dot{e}(s)\end{bmatrix}ds.$$

V_2 can be rewritten as $V_2=\frac{\tau(t)}{\tau_M}\varsigma^T P\varsigma+\frac{h_\tau(t)}{\tau_M}\int_{t-\tau(t)}^{t}\delta^T\frac{\Phi}{\tau(t)}\delta ds$. We can obtain $V_2>0$ if LMI (3) holds, then the LKF is positive. For any appropriately dimensioned matrix $U=[U_1^T\quad U_2^T]^T$, we have $0=2[e^T\quad e^T(t-\tau(t))]U[e-e(t-\tau(t))-\int_{t-\tau(t)}^{t}\dot{e}ds]$. For any appropriately dimensioned matrix V and scalar r, the following equation is true: $2[e^T+e^T(t-\tau(t))+r\dot{e}^T]V[\dot{e}-\dot{e}]=0$. Moreover, for any diagonal matrix $T_i>0$ ($i=1,2$), it follows from (2) that $\frac{\tau(t)}{\tau_M}g(1,t)+\frac{h_\tau(t)}{\tau_M}g(2,t)\geq 0$.

Case I. For $\tau(t)\in[\eta,\tau_C]$, we have $\dot{V}_2=\xi_1^T[Sym\{e_1\quad \eta e_9\quad 0\quad \tau_V e_{12}\}P\Psi^T]\xi_1+\gamma_1\xi_1^T[Sym\{0\quad 0\quad \tau_V e_{10}\quad 0\}P\Psi^T]\xi_1$ $\gamma_2\xi_1^T[Sym\{0\quad 0\quad \tau_V e_{11}\quad 0\}P\Psi^T]\xi_1-\xi_1^T\{[e_1\quad e_7]\overline{X}[e_1\quad e_7]^T+2h_\tau[e_1\quad e_7]\overline{X}[e_{13}\quad 0]^T\}$,

where $\gamma_1=\frac{\tau(t)-\eta}{\tau_V}\geq 0,\gamma_2=\frac{\tau_C-\tau(t)}{\tau_V}\geq 0,\gamma_1+\gamma_2=1$. From Lemmas 1 and 2, we have $V_5\leq\xi_1^T\{e_{13}[\eta^2 W_1$

$$+\tau_V^2(W_2+W_3)]e_{13}^T-[e_1\quad e_5\quad e_9]\Sigma_{W_1}[e_1\quad e_5\quad e_9]^T-[e_6\quad e_8\quad e_{12}]\Sigma_{W_3}[e_6\quad e_8\quad e_{12}]^T-Z_{W_2}\begin{bmatrix}\tilde{W}_2 & \overline{M}\\ & \tilde{W}_2\end{bmatrix}Z_{W_2}^T\}\xi_1.$$

Letting $L=VK$, it follows from above inequalities that $\dot{V}\leq\frac{h_\tau}{\tau_M}\xi_1^T\Theta_1^k\xi_1+\frac{1}{\tau_M}\int_{t-\tau(t)}^{t}\begin{bmatrix}\xi_1(t)\\ \dot{e}(s)\end{bmatrix}^T\Xi_1^k\begin{bmatrix}\xi_1(t)\\ \dot{e}(s)\end{bmatrix}ds$.

Therefore, LMIs (3)-(5) hold for $i=1,k=1,2$, which together with Schur complement equivalence imply that $\dot{V}<0$.

Case II. For $\tau(t)\in[\tau_C,\tau_M]$, by following a similar line of arguments as that in Case I, LMIs (3)-(5) hold for $i=2,k=1,2$, which together with Schur complement equivalence imply that $\dot{V}<0$.

Hence, it follows from the Lyapunov stability theory that the synchronization error system is absolutely stable, i.e., master system M and slave system S in (1) are global asymptotical synchronization from definition 1. This completes the proof.

4. Case study

Table 1. h with $\eta=0$.

Methods	h
[3]	0.3582
[6]	0.4355
[5]	0.4357
[8]	0.4369
[12]	0.4395
Theorem 1	0.6708

Table 2. h with $\eta=0$.

Methods	h
[13]	0.3914
[3]	0.3981
[14]	0.48
[5]	0.5147
[8]	0.5218
Theorem 1	0.7243

Table 3. h with different η.

η	Methods	r	h
0.1	[15]	0.63	0.22
	[5]	2	0.41
	Theorem 1	2	0.6635
0.2	[15]	0.63	0.064
	[5]	2	0.31
	Theorem 1	2	0.5894

Example [3,5,6,8,12]. Consider the following Chua's circuit (1) with sampled-data feedback control under the following parameters:

$$A=\begin{bmatrix}-am_1 & a & 0\\ 1 & -1 & 1\\ 0 & -b & 0\end{bmatrix},\ B=\begin{bmatrix}-c & 0 & 0\\ -c & 0 & 0\\ 2c & 0 & -c\end{bmatrix},\ H=\begin{bmatrix}-a(m_0-m_1)\\ 0\\ 0\end{bmatrix},\ C=D=[1\quad 0\quad 0].$$

In order to compare with some existing methods, let $d_\mu = 0$, i.e., $d(t) \equiv d$ is a constant. The maximum value of the sampling period h can be found in tables 1-3. We obtain the controller gain mat-

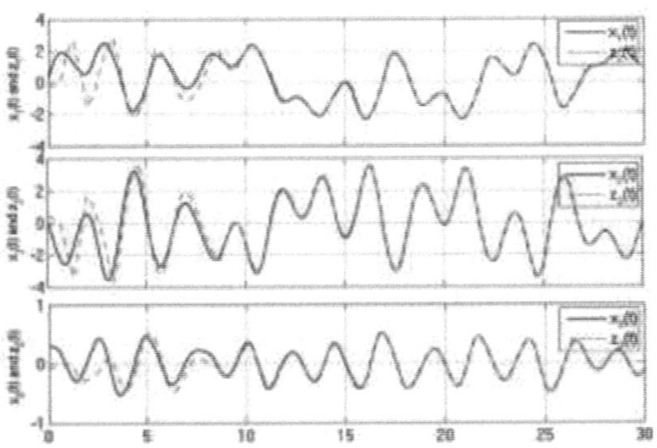

Figure 1. Response of x(t) and z(t) with h = 0.6708.

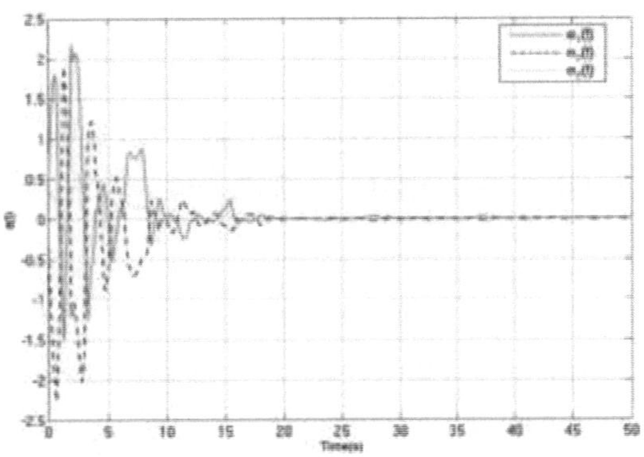

Figure 2. Response of e(t) with h = 0.6708.

-rix $K = [2.5597 \quad -0.1068 \quad -2.9233]^T$ with $h = 0.6708$ using Matlab LMI Toolbox. The responses of the states $x(t), z(t)$ and the error signal $e(t)$ of system (1) with the sampling period $h = 0.6708$ are given in figure 1 and figure 2.

5. Conclusions

In this paper, some new synchronization criteria are proposed for for time-delayed chaotic Lur'e systems using sampled data via a modified LKF combining the delay-fraction theory and the piecewise analysis method.

6. References

[1] L. Pecora and T. C. 1990 *Phys. Rev. Lett.*, **64** 821.
[2] Z. Wu, P. Shi, H. Su, and J. C. 2012 *IEEE Trans. Neural Netw. Learn. Syst.*, **23** 1368.
[3] Z. Wu, P. Shi, H. Su, and J. C. 2013 *IEEE Trans. Neural Netw. Learn. Syst.*, **24** 410.
[4] C. Ge, Z. Li, X. Huang, and C. S. 2014 *Nonlinear Dyn.*, **78** 2409.
[5] C. Ge, W. Zhang, W. Li, and X. S. 2015 *Neurocomputing*, **151** 215.
[6] C. Hua, C. Ge, and X. G. 2015 *IEEE Trans. Neural Netw. Learn. Syst.*, **26** 1214.
[7] K. Shi, X. Liu, H. Zhu, S. Zhong, Y. Liu, and C. Y. 2016 *Nonlinear Dyn.*, **83** 1259.
[8] K. Shi, X. Liu, H. Zhu, and S. Z. 2016 *Commun Nonlinear Sci. Numer. Simul.*, **34** 165.
[9] Z. Liu, L. Yu, and D. X. 2013 *Commun. Nonlinear Sci. Numer. Simul.*, **18** 1246.
[10] A. Seuret and F. G. 2013 *Automatica*, **49** 2860.
[11] Y. Liu, S. Lee, O. Kwon, and J. P. 2015 *Circuits Syst. Signal Process*, **34** 1481.
[12] J. Cao, R. Sivasamy, and R. R. 2016 *Circuits Syst. Signal Process*, **35** 811.
[13] W. Chen, Z. Wang, and X. L. 2012 *IEEE Trans. Circuits Syst. II: Express Briefs*, **59** 515.
[14] C. Zhang, et al. 2013 *Commun. Nonlinear Sci. Numer. Simulat.*, **18** 2743.
[15] S. Theesar, S. Banerjee, and P. B. 2012 *Nonlinear Dyn.*, **70** 1977.

Acknowledgments
This work is supported partially by the National Natural Science Foundation of China under Grant nos. 61603325 and 61573188, Natural Science Foundation of Jiangsu Province under Grant no. BK20160441, Natural Science Foundation for Colleges and Universities of Jiangsu Province under Grant no. 16KJD120002 and the Talent Introduction Project of YanCheng Institute of Technology under Grant no. xj201516.

IOP Conf. Series: Journal of Physics: Conf. Series **814** (2017) 012023 doi:10.1088/1742-6596/814/1/012023

Efficient parallel analysis of transient plane wave scattering from nonplanar doubly periodic structures

Nan-Wei Chen

Department of Communications Engineering, Yuan Ze University, Jhongli 320, Taiwan

nwchen@saturn.yzu.edu.tw

Abstract. This paper presents a fast scheme for solving a time-domain electric field integral equation pertinent to the analysis of transient plane wave scattering from nonplanar doubly periodic structures. The proposed scheme efficiently evaluates scattered fields from periodic structures by decomposing the sources that produce them into present and past components. The fields radiated by present sources are evaluated classically. In contrast, fields generated by past sources are expanded into time domain Floquet waves with modal amplitudes that are evolved in both space and time using a spectral time-stepping scheme complemented by Huygens' principle based boundary conditions. The kernel equations and efficient parallel evaluation associated with the scattered fields generated by past sources are validated and effectiveness of the resulting algorithm are demonstrated by a number of examples.

1. Introduction

In the past, transient plane wave scattering from doubly periodic structures has been analyzed predominately using finite difference methods [1-3]. Classically formulated finite difference schemes however cannot be applied directly to the analysis of scattering from obliquely illuminated periodic structures as they require future field values to update present ones. While several avenues for resolving this problem have been proposed [4-6], most are hard to implement and/or somewhat limited in scope. The integral equation based solvers for analyzing transient plane wave scattering form doubly periodic perfectly electrically conducting (PEC) free-standing [7] or substrate imprinted PEC elements [8] have been proposed in the past. The solvers [7][8] evaluate so-called scattered fields produced by transient periodic current constellations by using free-space Green functions, and the evaluation of the fields on/in the scatterer has to be carried out for each time step in the analysis. Hence, the solvers are rendered computationally expensive and inapplicable to the analysis of real-world structures. Indeed, it is shown that their computational complexity scales as $O\left(N_s^2 N_t^3\right)$ where N_s denotes the number of spatial unknowns associated with currents in one of the structure's cells – also termed the reference or mother-cell – and N_t is the number of time steps in the analysis.

To allievate this computational burden, the periodic structure solver [9] was presented and the burden was alleviated by expanding scattered fields into time domain Floquet waves (TDFWs) [10] but is limited to discretely planar structures, viz. structures comprising of a finite number of planar screens. This solver exploits the fact that fields produced by periodic and transient but quiescent current constellations can be expanded in terms of a finite number of propagating TDFWs in order to rapidly evaluate scattered fields by splitting them into two components. First, there are the fields radiated by the sources in the mothercell and its immediate vicinity; these sources are termed *present*, and their fields are evaluated classically. Second, there are the fields generated by sources that are spatially removed from the mothercell; these sources are termed *past*, and their fields are expanded in TDFWs. Unfortunately, when sources and observers reside on different planes, the TDFW propagator no longer is time shift invariant; this fact in itself renders this technology inapplicable to the analysis of nonplanar structures. Here, to render TDFW-based TDIE solvers applicable to the analysis of transient plane wave scattering from nonplanar periodic structures, a new scheme for tracking TDFW amplitudes is proposed. The scheme evolves TDFW modal amplitudes in (one-dimensional) space and time using a spectral time-stepping scheme supplemented with Huygens' principle based boundary conditions. Because the scheme in no way requires the TDFW propagator to be time shift invariant, it applies without difficulty to the analysis of nonplanar doubly periodic structures consisting of PEC elements and sculptured dielectric volumes, including that of PEC elements imprinted on nonplanar substrates. Furthermore, the TDFW modal amplitudes can be parallelly evaluated.

This paper is organized as follows. In Section 2, the development of the proposed Floquet-wave-based solver for doubly periodic structures described is described. Numerical results that demonstrate the capabilities and accuracy of the proposed method are presented in Section 3. Finally, Section 4 draws the conclusions.

2. Problem statement and formulation

Consider a doubly periodic structure consisting of PEC surfaces S and penetrable volumes V. These volumes comprise isotropic though potentially inhomogeneous dielectrics with frequency independent permittivity $\varepsilon(\mathbf{r})$ (figure 1). The periodic arrangement $S \cup V$ resides in free-space with permittivity ε_0. Materials occupying V are assumed nonmagnetic, i.e., of free space permeability μ_0. As illustrated in figure 1, the surfaces S and penetrable volumes V are decomposed as $S = \bigcup_{m,n=-\infty}^{+\infty} S_{mn}$ and $V = \bigcup_{m,n=-\infty}^{+\infty} V_{mn}$, with S_{mn} and V_{mn}, $m,n = -\infty,+\infty$, termed sub-surfaces and –volumes residing in cell (m,n) anchored to vector $\mathbf{r}_{mn}^c = mD_x\hat{\mathbf{x}} + nD_y\hat{\mathbf{y}}$, where D_x and D_y denote unit cell dimensions along x and y directions, respectively; S_{00} and V_{00} are said to reside in the mother cell. The structure is illuminated by a bandlimited plane wave pulse propagating along direction $\hat{\mathbf{k}}^{inc} = -\sin\theta^{inc}\cos\phi^{inc}\hat{\mathbf{x}} - \sin\theta^{inc}\sin\phi^{inc}\hat{\mathbf{y}} - \cos\theta^{inc}\hat{\mathbf{z}}$ with electric field $\mathbf{E}^{inc}(\mathbf{r},t,\hat{\mathbf{p}}^{inc}) = \hat{\mathbf{p}}^{inc} f(t - \hat{\mathbf{k}}^{inc}\cdot\mathbf{r}/c)$ where c is the speed of light in free-space and $\hat{\mathbf{p}}^{inc}$ denotes the incident field's polarization. It is assumed that $\mathbf{E}^{inc}(\mathbf{r},t,\hat{\mathbf{p}}^{inc})$ is bandlimited to $\omega = \omega_{\max}$ and vanishingly small on S_{00} and V_{00} for $t < 0$. In response to the above excitation, surface currents $\mathbf{J}_{mn}^s(\mathbf{r},t) = \mathbf{J}_{00}^s(\mathbf{r} - \mathbf{r}_{mn}^c, t - \hat{\mathbf{k}}^{inc}\cdot\mathbf{r}_{mn}^c/c)$ flow on S_{mn} and volumetric polarization currents $\mathbf{J}_{mn}^v(\mathbf{r},t) = \mathbf{J}_{00}^v(\mathbf{r} - \mathbf{r}_{mn}^c, t - \hat{\mathbf{k}}^{inc}\cdot\mathbf{r}_{mn}^c/c)$ arise in V_{mn}. Note that $\mathbf{J}_{00}^v(\mathbf{r},t) = \kappa(\mathbf{r})\frac{\partial}{\partial t}\mathbf{D}_{00}^{tot}(\mathbf{r},t)$, where $\kappa(\mathbf{r}) = (\varepsilon(\mathbf{r}) - \varepsilon_0)/\varepsilon(\mathbf{r})$ denotes the contrast ratio.

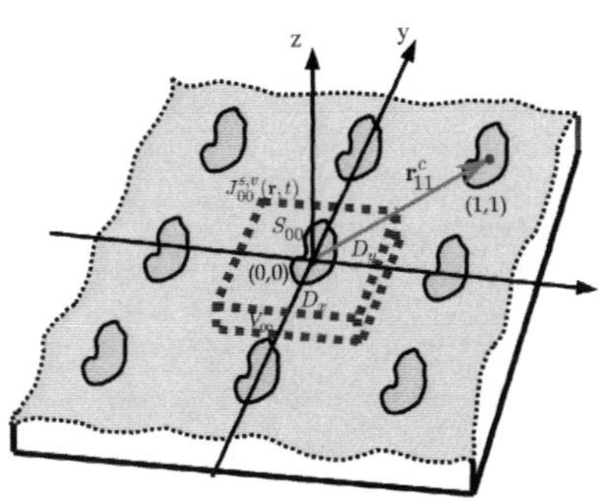

Figure 1. Illustration of the doubly periodic structure.

As detailed in [10], enforcing the (temporal derivative of the) total electric field tangential to S_{00} to vanish and imposing a consistency condition on the total electric field within V_{00} yields the coupled electric field integral equations for $\mathbf{J}_{00}^{s}(\mathbf{r},t)$ and $\mathbf{D}_{00}^{v}(\mathbf{r},t)$. The scattered fields from $\mathbf{J}_{00}^{s}(\mathbf{r},t)$ and $\mathbf{D}_{00}^{v}(\mathbf{r},t)$ are expresses as

$$\frac{\partial}{\partial t}\mathbf{E}^{sca,s}(\mathbf{r},t) = -\frac{\mu_0}{4\pi}\frac{\partial^2}{\partial t^2}\int_{S_{00}} ds'\mathbf{J}_{00}^{s}(\mathbf{r}',t) * G(\mathbf{r},\mathbf{r}',t) + \frac{1}{4\pi\varepsilon_0}\nabla\int_{S_{00}} ds'\left[\nabla'\cdot\mathbf{J}_{00}^{s}(\mathbf{r}',t)\right] * G(\mathbf{r},\mathbf{r}',t), \tag{1}$$

and

$$\frac{\partial}{\partial t}\mathbf{E}^{sca,v}(\mathbf{r},t) = -\frac{\mu_0}{4\pi}\frac{\partial^2}{\partial t^2}\int_{V_{00}} dv'\kappa(\mathbf{r}')\frac{\partial\mathbf{D}_{00}^{tot}(\mathbf{r}',t)}{\partial t} * G(\mathbf{r},\mathbf{r}',t) + \frac{1}{4\pi\varepsilon_0}\nabla\int_{V_{00}} dv'\left[\nabla'\cdot\kappa(\mathbf{r}')\frac{\partial\mathbf{D}_{00}^{tot}(\mathbf{r}',t)}{\partial t}\right] * G(\mathbf{r},\mathbf{r}',t), \tag{2}$$

respectively. That is, the evaluations of the scattered field associated with $\mathbf{J}_{00}^{s}(\mathbf{r},t)$ and $\mathbf{J}_{00}^{v}(\mathbf{r},t)$ rely on the following Green's function for the doubly periodic structure,

$$G(\mathbf{r},\mathbf{r}',t) = \sum_{m=-\infty}^{m=+\infty}\sum_{n=-\infty}^{n=+\infty}\frac{\delta(t-\hat{\mathbf{k}}^{inc}\cdot\mathbf{r}_{mn}^{c}/c-\left|\mathbf{r}'+\mathbf{r}_{mn}^{c}-\mathbf{r}\right|/c)}{\left|\mathbf{r}'+\mathbf{r}_{mn}^{c}-\mathbf{r}\right|}. \tag{3}$$

The coupled electric field equations are solved with a marching-on-in-time procedure via approximating S_{00} and V_{00} with planar triangular facets and tetrahedral volumes, respectively. Also, Rao-Wilton-Glisson functions [11] and their volumetric extensions proposed by Schaubert et al. [12] are chosen to represent the spatial variations of $\mathbf{J}_{00}^{s}(\mathbf{r},t)$ and $\mathbf{D}_{00}^{v}(\mathbf{r},t)$. And, the approximate prolate spheroidal wave functions [13] are adopted to represent the temporal variations of $\mathbf{J}_{00}^{s}(\mathbf{r},t)$ and $\mathbf{D}_{00}^{v}(\mathbf{r},t)$. Unfortunately, the double summation in (3) leads to a computational burden. However, this burden can be drastically relieved through the introduction of the concept of TDFWs and the use of a spectral stepping scheme complemented by Huygens' principle based boundary conditions. In essence, the Green's function can be expanded in terms of TDFWs [10] as

$$G(\mathbf{r},\mathbf{r}',t) = \sum_{p=0}^{\infty}\sum_{q=-\infty}^{\infty}\beta_{pq}\Re e\left[A_{pq}^{FW}(\mathbf{r},\mathbf{r}',t)\right] \tag{4}$$

where $\beta_{pq}=1$ for $p=0$ and $\beta_{pq}=2$ for $p\neq 0$, $\Re e[\cdot]$ selects the real part,

$$A_{pq}^{FW}(\mathbf{r},\mathbf{r}',t) = \frac{ce^{-j\boldsymbol{\alpha}_{pq}\cdot(\boldsymbol{\rho}-\boldsymbol{\rho}')}}{2D_xD_y\sqrt{1-\eta^2}}e^{j\bar{\omega}_{pq}\tau}J_0[\tilde{\omega}_{pq}\sqrt{\tau^2-\tau_0^2}]U(\tau-\tau_0), \tag{5}$$

$$\begin{aligned}
&\mathbf{r} = \boldsymbol{\rho} + z\hat{\mathbf{z}} && \bar{\omega}_{pq} = \frac{\eta c}{1-\eta^2}\hat{\mathbf{u}}_1\cdot\boldsymbol{\alpha}_{pq} \\
&\mathbf{r}' = \boldsymbol{\rho}' + z'\hat{\mathbf{z}} && \tilde{\omega}_{pq}^2 = \bar{\omega}_{pq}^2 + \alpha_{pq}^2c^2/(1-\eta^2) \\
&\eta = \sin\theta^{inc} && \tau_0 = \frac{\sqrt{1-\eta^2}(z-z')}{c} \\
&\boldsymbol{\alpha}_{pq} = 2\pi p/D_x\,\hat{\mathbf{x}} + 2\pi q/D_y\,\hat{\mathbf{y}} && \tau = t - t_0 \\
&\alpha_{pq}^2 = (2\pi p/D_x)^2 + (2\pi q/D_y)^2 && t_0 = \frac{\eta\hat{\mathbf{u}}_1\cdot(\boldsymbol{\rho}-\boldsymbol{\rho}')}{c} \\
&\hat{\mathbf{u}}_1 = \cos\phi^{inc}\hat{\mathbf{x}} + \sin\phi^{inc}\hat{\mathbf{y}} &&
\end{aligned} \tag{6}$$

IOP Conf. Series: Journal of Physics: Conf. Series **814** (2017) 012023 doi:10.1088/1742-6596/814/1/012023

$j = \sqrt{-1}$, $J_0(\cdot)$ denotes the zeroth order Bessel function of the first kind and $U(\cdot)$ is the Heaviside step function. Although the summation in (4) comprises an infinite number of terms, upon convolution of the Green function with a bandlimited and essentially time limited signal, only finite number of so-called propagating modes should be retained provided that the field is observed no earlier than $-(\mathbf{r}-\mathbf{r}')\cdot\hat{\mathbf{k}}^{inc}/c$ seconds after the signal (essentially) vanishes in the temporal dimension.

Hence, the fields are evaluted via a space-time convoltion of the Green function and the current sources on/in the mothercell. More specific, the fields at a given time can be expressed in terms of instantaneous and delayed field, which are respectively attributed to the present and past currents on the mothercell. The evaluation of the instantaneous fields is carried out classically with (1), while the delayed fields are computed by expanding them into a finite number of TDFWs. In essence, the fast scheme bears significant similarity to the plane wave time domain algorithm [14]. It appears that fields generated by quiescent and bandlimited periodic currents can be represented in terms of a few propagating TDFWs. A four-stage scheme is developed to evolve the amplitude of each propagating TDFW. The TDFWs are propagated like plane waves in the plane of the periodicity and are easy to manipulate. However in the direction perpendicular to the periodicity, the TDFWs are not time-shift invariant anymore. A spectral time-stepping algorithm supplemented with Huygens' principle based boundary conditions is designed to efficiently track the TDFW modal amplitudes in the vertical direction. On the other hand, the computation of the delayed fields can be further accelerated with parallel computing owing to the TDFW modal nature.

3. Numerical results

The proposed scheme is exploited to analyze transient scattering from the following doubly periodic structures. The results obtained were compared against data from a periodic frequency domain method of moments code (P-MOM) following Fourier transformation of the time-domain currents. All the examples presented below are illuminated by a modulated Gaussian pulse parameterized by

$$f(t) = \cos\left[2\pi f_c(t-t_p)\right]\exp\left[-(t-t_p)^2/(2\sigma^2)\right] \tag{7}$$

where f_c is center frequency of the incident wave, $\sigma = 3/(2\pi f_{bw})$ and $t_p = 6\sigma$ with f_{bw} termed the "bandwidth" of the signal.

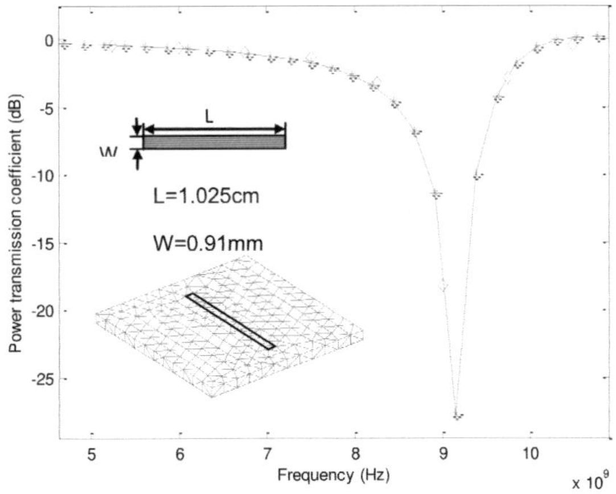

Figure 2. Power transmission coefficient plot of the doubly periodic dipole structure.

The first example analyzed is the periodically imprinted dipole elements shown in the inset of figure 2. The dielectric slab has a relative permittivity $\varepsilon_r = 3.0$ and thickness 1 mm. The side length of the square mother cell is 1.78 cm. The PEC dipole is of dimension 1.025 cm by 0.91 mm. The structure is discretized in terms of 2656 spatial unknowns. The incident field has $\hat{\mathbf{p}}^{inc} = \hat{\mathbf{y}}$, f_c=6 GHz , and f_{bw} =6 GHz. The time step is Δt=4.167 ps and 512 time steps are used in the analysis. The number of Floquet modes used for field expansion is 81. The power transmission coefficient of the structure obtained via the Floquet-wave-based scheme and P-MOM scheme are compared in figure 2. The good agreement between the two data sets is observed.

The second example analyzed is the periodically imprinted dual-loop elements. The dielectric slab has a relative permittivity $\varepsilon_r = 3.0$ and thickness 0.6 cm. The side length of the mother cell is 6 cm and the dimension of the dual-loop is shown in the inset of figure 3. The structure is discretized in terms of 2398 spatial unknowns**.** The incident field has $\hat{\mathbf{p}}^{inc} = \hat{\mathbf{x}}$, f_c=1.7 GHz and f_{bw}=1.7 GHz. The time step is Δt=14.706 ps and 1024 time steps are used. The number of Floquet modes used for field expansion is 73. The power transmission coefficient of the structure obtained using the proposed scheme and P-MOM scheme are compared in figure 3. The results are in good agreement.

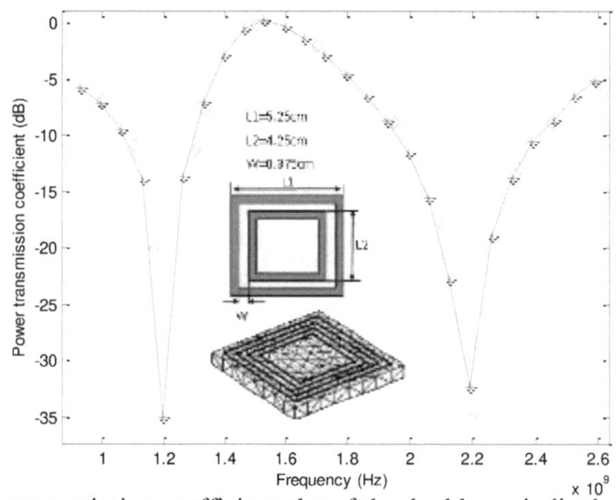

Figure 3. Power transmission coefficient plot of the doubly periodic dual-loop structure.

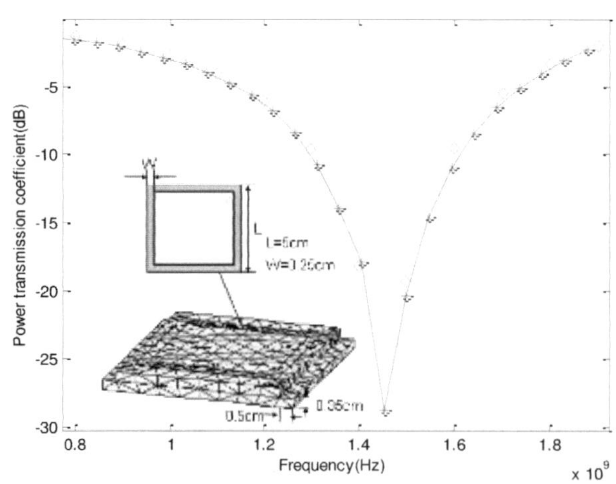

Figure 4. Power transmission coefficient plot of the grooved doubly periodic square-loop structure.

The next example analyzed is the periodically arranged square-loop elements imprinted on a grooved dielectric slab. The grooved dielectric slab has a relative permittivity $\varepsilon_r = 3.0$. The thicknesses are 0.7 cm and 0.35 cm for the un-grooved and grooved sections, respectively. The side length of the square mother cell is 7 cm. The dimension of the square-loop is shown in the inset of figure 4. The geometry is discretized in terms of 1908 spatial unknowns. The incident field has $\hat{\mathbf{p}}^{inc} = \hat{\mathbf{x}}$, f_c=1.2 GHz and f_{bw}=1.2GHz. The time step is Δt=20.83 ps and 1024 time steps are used. The number of Floquet modes used for field expansion is 49. The power transmission coefficient of the structure obtained via the proposed scheme and P-MOM scheme are compared in figure 4. Again, the good agreement between the two data sets is observed.

4. Conclusions

An improved Floquet-wave-based scheme for analyzing transient scattering from nonplanar doubly periodic structures is described. It relies on evolving individual TDFWs using a spectral time-stepping scheme supplemented with Huygens' principle based boundary conditions. Also, the modal field evaluations are accelerated by parallel computing. The scheme applies without difficulty to nonplanar doubly periodic structures comprising PEC and sculptured dielectric substrates. The efficiency and accuracy of the proposed scheme have been demonstrated via comparison against frequency domain results. Detailed formulations together with the implementations on the acceleration will be presented at the conference.

5. References

[1] J. A. Roden, S. D. Gedeny, M. P. Kesler, J. G. Maloney, and P. H. Harms 1998, *IEEE Transactions on Microwave Theory and Technology*, **46**
[2] A. Y.-C. Kao and R. G. Atkins 1996, *IEEE Antenna and Propagation Society Intl. Symp*
[3] W. J. Tsay and D. M. Pozar 1993, *IEEE Microwave Guided Wave Letter*, **3**
[4] M. E. Veysoglu, R. T. Shin, and J. A. Kong 1993, *Journal of Electromagnetic Waves and Applications*, **7**
[5] H. Holter and H. Steyskal 2002, *IEEE Transactions on Antennas and Propagation*, **47**
[6] J. G. Maloney and M. P. Kesler, *Computational Electrodynamics: the finite-difference time-domain* (A. Taflove and S. C. Hagness, Artech House, Boston, 2002)
[7] N.-W. Chen, B. Shanker, and E. Michielssen 2003, *IEE Proceedings- Microwave, Antennas and Propagation*, **150, 2**
[8] N.-W. Chen, B. Shanker, and E. Michielssen 2002, *IEEEAPS Intl. Symp.*
[9] N.-W. Chen, M. Lu, B. Shanker, E. Michiellsen, and F. Capolino 2003, *URSI North American Radio Science Meeting*, Columbus, OH
[10] F. Capolino and L. Felsen 2000, *IEEE Trans. Antennas Propagat.*, **48,6**
[11] S. M. Rao, D. R. Wilton, and A. W. Glisson 1982, *IEEE Transactions on Antennas and Propagation*, **30,3**
[12] D. H. Schaubert, D. R. Wilton, and A. W. Glisson 1984, *IEEE Transactions on Antennas and Propagation*, **32,1**
[13] J. J. Knab1979, *IEEE Transactions on Information Theory*, **25, 6**
[14] A. A. Ergin, B. Shanker, and E. Michielssen 1998, *Journal of Computational Physics*, **146**

Institute of Physics
Dirac House, Temple Back
Bristol BS1 6BE UK

ISSN: 1742-6588
ISBN 978-1-5108-3859-8

2nd International Conference on Advanced Materials Research and Manufacturing Technologies (AMRMT 2017)

IOP Conference Series: Materials Science and Engineering
Volume 229

Phuket, Thailand
2 – 5 August 2017

IOP
Institute of Physics